# The view from space
## PHOTOGRAPHIC EXPLORATION OF THE PLANETS

**MERTON E. DAVIES**

**BRUCE C. MURRAY**

Columbia University Press  New York and London  1971

# The view from space
## PHOTOGRAPHIC EXPLORATION OF THE PLANETS

*Merton E. Davies* is with The Rand Corporation.

*Bruce C. Murray* is Professor of Planetary
Science at the California Institute
of Technology.

# Foreword

The Rand Corporation has had a long and continuing interest in spacecraft, especially in their use as observation platforms. It therefore seems appropriate for Rand to sponsor this book, which treats the problems and potentials of photography of the Earth, Moon, and planets from space.

Rand's interest in space began in late 1946, when it performed initial analyses of satellite feasibility for the Air Force. These and subsequent studies were prompted by the realization after World War II that the wartime rocket technology had enormous implications for peacetime space activities. It was concluded that the major role of military satellites should be for observation rather than aggressive warfare.

Responsibility for the U.S. satellite program was assigned to the Air Force in 1948, and Rand was asked by the Air Force to manage this effort. By 1951 the political implications of satellites and their over-all use in the interests of science had been examined, in addition to the major focus on mission analysis and design. In the same year, Rand published an important report on using satellites for weather observation.

In the early 1950s, when the Air Force started to develop actual satellite systems, Rand served as the advisor for this expanded effort. Rand research efforts contributed as well to a number of vehicle developments and space systems, particularly the Discoverer recoverable satellite and the Agena launch vehicle sys-

tems, programs which have evolved to permeate today's entire U.S. space program.

In the middle 1950s, Rand expanded its interests to include lunar and planetary vehicles as well as satellites for scientific exploration. Major Rand studies included a series of reports on a lunar instrument carrier in 1956. Another study led to very intensive work on photography from space—work which culminated in a public patent on a specific system for space photography in 1959, issued to Merton Davies, one of the authors of this book.

Also in 1959, Rand compiled an authoritative introduction to space activities in response to widespread public and congressional interest in space activities. This book, *The Space Handbook*, was widely disseminated to bring about a more enlightened perspective on the rapidly growing space program.

In 1960 a well-known Rand researcher, Amrom Katz, wrote a fundamental series of articles on observation satellites which ran in six consecutive issues of *Astronautics* and gave special prominence to space photography. These articles by Katz are still one of the best introductions to the problems of designing photographic systems to yield information from space platforms.

Political interest in the role of observation satellites was heightened as a result of the meetings at Geneva between the Western and Eastern blocs during November and December of 1958, when the role of satellites was explored to consider technical aspects of preventing surprise attack. Mr. Davies was a member of the U.S. delegation to that conference. These rather detailed studies were published in January 1959 and contain the first significant public references to the use of satellite photography to monitor arms build-ups and to prevent surprise attack, a subject given further consideration in the present book.

Many of the possibilities foreseen in these documents of the 1950s have been pursued in the last decade, although the results in some instances are necessarily obscured by the requirements of national security. A major Rand interest, and a special interest of the two authors of this book, is the nonmilitary use of observations from space.

Space photography has many practical applications. Satellites which gather weather information for both military and public purposes have been in use for a decade; surveillance satellites designed to discover, observe, and measure the earth resources useful to man are of much current interest. Indeed, satellite observation seems destined to become especially significant for earth resource applications, with a function comparable in importance to that of communications satellites.

Early space photography was severely constrained by optical and spacecraft limitations. However, the unique potentialities of space photography spurred extraordinary technological developments so that initial constraints generally have been overcome. Most of the current limitations are economic rather than

technological. Furthermore, present technological capability permits the reliable prediction of potential performance; thus, proposed future objectives rest on a firm technological foundation.

This book recounts past history and contemporary achievements. But beyond an exposition of the results which flow from space photography and their impact on contemporary scientific theories, the author's aim is to show the impact of these results on men's minds generally. The authors are unusually well qualified because of the breadth of their past interests. They have been involved in theoretical analyses and studies of potential applications of space photography, and in the associated problems of balancing technical capabilities with very difficult experimental demands. Furthermore, they have been active participants in major space experiments involving space photography.

Merton Davies, a senior staff member of The Rand Corporation, participated in the early Rand work on satellites and space vehicle technology and has for the last several years devoted much of his attention to planetary photography. A co-investigator in the television experiments of the Mariner 6 and 7 probes, Mr. Davies is serving in the same capacity for the 1971 orbiting Mars mission and the Venus/ Mercury flyby mission in 1973.

Dr. Bruce Murray, a Rand consultant since 1961, is Professor of Planetary Science at the California Institute of Technology, and is an expert on lunar and planetary surfaces. A co-investigator in the television experiment of the Mariner 4, 6 and 7 flybys of Mars, Professor Murray is also a co-investigator of the television team for 1971 orbiting experiments. He is head of the Imaging Team for the 1973 Mercury/Venus mission.

Both men have done much in adapting the technology developed in the early phases of the U.S. space program to planetary photography. It is from this wealth of contemporary experience that the two authors discuss the problems, rewards, and implications of space photography.

Planetary photography from space vehicles is one of the supremely challenging and exciting aspects of our current space program. It is a unique and not-to-be-repeated experience to survey distant objects in the solar system from a close-hand perspective and to be able to record those surveys in permanent records of great historical significance. The excitement of these experiments is that they record singular events in man's first new perceptions of his external universe. That excitement is one of the rewarding experiences revealed by this book.

In nature's infinite book of secrecy
A little I can read
*Antony and Cleopatra* i.ii.11

The Rand Corporation, Santa Monica, California

Bruno Augenstein
Vice President

# Preface

Space is intrinsically remote and impersonal. Yet the interaction of the human mind and spirit with space can be intensely personal and intimate and truly constitutes a unique aspect of the twentieth-century human experience. This book is an attempt to communicate widely the visions, difficulties, hopes, and frustrations which have characterized the first decade of space exploration by photography and to use this collective experience as a basis for charting a course through the next decade and beyond.

The view *of* space presented in the text, unlike the photographic views *from* space presented in the illustrations, is colored and limited by the experiences, imagination, and prejudices of its authors. For this we can offer no apologies since omniscience and objectivity are not found in any greater abundance among space scientists—or any other group of scientists—than elsewhere in the population. If we manage to communicate some of our views as participants in space exploration to members of the larger community who find themselves cast as willing or even involuntary spectators, then our efforts will have been warranted. The book should also be of value in a more scholarly way to those interested in the history of certain parts of the space program or in the objectives and techniques of space photography.

Although we are experienced professionals in space endeavors, we began this literary venture in blissful ignorance of the time and effort which would ultimately be required to say what we have to say. Consequently, the book was in preparation for more than two years. It is current through the end of 1970, and there has been minor editing through the first part of 1971. In a few instances, material published previously has been substantially rewritten and updated for use here. But, generally, completely new material has been written, reviewed by professional colleagues, and then rewritten. In that regard, we are indebted to a large number of our colleagues who have given so generously of their time in the unglamorous task of technical review and who may, in some instances, feel that their advice is not entirely apparent in the final product. We acknowledge with sincere appreciation the criticisms and suggestions of:

Mr. William B. Graham, Mr. E. Charles Heffern, Dr. Malcolm Hoag, Mr. Francis Hoeber, Mr. Ralph Hushke, and Mr. Amrom Katz of Rand.

Dr. Roger Bourke, Mr. James Burke, Mr. Edward Danielson, Dr. Phillip Eckman, Dr. James Long, Dr. Robert Mackin, and Dr. Donald Rea of JPL.

Professors Norman Horowitz and Robert Leighton of Caltech, and Professor Carl Sagan of Cornell University.

Dr. Newell Trask of the United States Geological Survey, Mr. Donald Gault of the NASA Ames Research Center, and Mr. Bradford Smith of New Mexico State University.

In addition we wish to express our appreciation of the cheerful and talented efforts of

ix

Jurrie van der Woude and Ruth Talovitch of Caltech in photography and illustration, of James Anderson and Gordon Hoover of Caltech in computation, and of Jeanne Dunn and William Taylor of Rand in editorial criticism.

Finally, a word of appreciation for The Rand Corporation and the California Institute of Technology. Without the intellectual stimulation and freedom characteristic of both institutions there would have been little opportunity for us to develop a point of view worth writing a book about. And, without the generous financial support and encouragement by Rand throughout the extended period of preparation, it would not have been possible to have produced a book in the midst of the intense and continuing space involvements of both of us.

It should be apparent that the views presented in this book are the sole responsibility of the authors and do not imply endorsement by any institution or other persons.

*February 1971*

Merton E. Davies
*Santa Monica*

Bruce C. Murray
*Pasadena*

# Contents

# The view from space
## PHOTOGRAPHIC EXPLORATION OF THE PLANETS

VOYAGE A LA LUNE

# CHAPTER 1
## Introduction

### 1.1 THE CULTURAL SIGNIFICANCE OF SPACE PHOTOGRAPHY

During the last decade, man progressed from his first brief flight into space to landing and walking on the Moon itself. But even while United States and Soviet astronauts were first testing the Earth-orbital environment, other explorers were already peering at Mars through the robot eyes of Mariners and remotely examining the Moon at close-hand with Rangers, Surveyors, and Orbiters. The ancient dream of space exploration no longer belongs solely to the visionaries. Practical men in our government now debate seriously which decade should be our goal for sending man himself to the planets. Most agree that the extraordinary first successes of the probes to Mars and Venus should be followed promptly by the sending of picture-taking robots to all the planets as well as the practical exploitation of Earth's orbit.

Popular enthusiasm for space exploration, however, is not a sufficient reason in itself. There must be a more thoughtful answer to the question, "Why space?" The American public correctly expects justification of the large expenditures of money and highly skilled people which would be needed in the future. Indeed, there is a close association in the public's mind between "Space" and "Defense" because they share a common industrial base. It must be made clear to what extent our scientific program to explore the planets, the Moon, or even the Earth itself stems from cold war competition and how much from more positive cultural aspirations.

There are, of course, tangible benefits from space already—communication and weather satellites, for example—and many others on the drawing board. In this book we will contend that the real justification for many kinds of space activities lies in their impact—through photography—upon the minds of men. Whether it be the new perspective of a color view from Earth orbit showing the deserts of the troubled Middle East, or the drama of Armstrong's "First Small Step" on live television, or the wonder of a Mariner view of the polar cap of Mars depicting a landscape never before seen nor imagined, the view from space changes the consciousness of every viewer. The Earth photographed from space (see Figure 1.1) lacks the national boundary lines separating the neatly colored areas of our maps; man must adjust—and grow—as he realizes that the *colored map* is the artificial representation and that the photograph more correctly depicts reality. The image of Armstrong stepping out onto the lunar surface (see Figure 1.2) permitted hundreds of millions of people to step out with him. The Moon, all of a sudden, became a more intimate place and geocentrism dwindled accordingly. The strangeness of the Martian lanscape (see Figure 1.3) feeds the human appetite for novelty with reality instead of fantasy, reminding us of the diversity, richness, and extension of the physical universe of which we are a part.

A black-and-white print of the original color photograph
taken by U.S. astronauts in September 1966 from
Gemini XI. The triangularly-shaped Sinai Desert
forms the center of the view, bounded on the lower left
(west) by the Gulf of Suez and the Suez Canal. The
Gulf of Aquaba, the Dead Sea, and the distant Sea of
Galilee are linearly-aligned to the upper left. Across
these natural boundaries the affairs of Israel, Jordan,
Syria, Iraq, and Egypt continue their tumultuous and
often tragic course. Much of Jewish, Moslem, and
early Christian history transpired within the area shown
by this photograph. Of ecological interest is the
black smoke to the far right originating from an oil well
fire in Iraq.

Space photography is a process of collective
discovery, not just by scientists alone but by all
men. The specialist, in fact, may eventually
become so familiar with his view from space
that he ceases to react to that view. But to the
human race as a whole, diverse views of the
Earth, the Moon, and the planets will represent
an increasingly important part of the remem-
bered images common to everyone. Their
impact on man's view of himself, of his nation
and his planet, may in the long run be comparable to the nineteenth century's annihilation
of a man-centered view of time. Just as a
child's early visualization of his surroundings
profoundly affects his adult concept of reality,
so may man's visualization of his whole Earth

and his Solar System ultimately affect man's
view of Man.

The authors of this book feel that exploration
is a basic human activity of lasting cultural
value. Even though large-scale exploration has
always been characterized by mixed motives—
economic exploitation, national prestige, genuine curiosity—the net result has usually been
of enduring cultural significance. If the cultural
score card could be totaled up for each century,
geographic discoveries would be among the
most valued products. And since space photography now provides such capability for discovery and also for its wide dissemination,
this application of the technology has unique
cultural value.

We live in turbulent times and are sometimes disheartened by ever-increasing technological capability of implementing man's
inhumanity to man and by traumatic changes
in our social and physical environments resulting from industrialization. But such are the
consequences of man's fateful evolution as a
tool-maker. None of us chose to be born in the
twentieth century. One of the fruits of this
accidental circumstance is the opportunity for
positive achievement beyond the dreams of our
predecessors. There is a unique advantage to
being a twentieth-century American, or Russian. We can, if we but choose, discover things
that could never be discovered before, see
worlds that have never been seen before, and
moreover now make that view from space part
of the vision of all men.

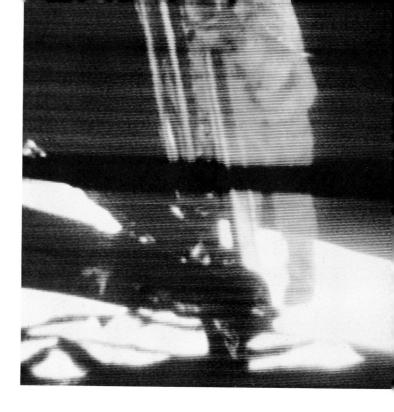

### 1.2 Armstrong's "first small step"

The television view seen live by an estimated 500 million people throughout the world as astronaut Neil Armstrong first set foot on the Moon on July 20, 1969. Although of very low quality as a picture, this image is (or was) probably one of the images most common to all the world's peoples and thus has significance in earthly social terms as well as a symbol of manned lunar exploration.

### 1.2 THE ANATOMY OF SPACE

Photography is not a space age development; there has been a continuing development of its technology ever since its inception in the nineteenth century. The advent of balloons, then airplanes, and finally rockets provided ever higher platforms from which to view the Earth. Much of the technological background for space photography was brought into being by the rapid development of aerial photography during and after World War II. Thus it is space, not photography, that is most new, and the unique techniques and results of space photography derive from the opportunities and difficulties peculiar to the space environment.

Similarly, the role and significance of space photography need to be viewed within the context of the why's and wherefore's of man's first venture into space from his terrestrial home. An inquiry into the nature of the first decade of space flight can be conveniently made in terms of the U.S. versus the USSR, exploration versus exploitation, and manned versus unmanned techniques. This format is followed here. Further, we will for the moment try to gain perspective on space by looking at space activities in somewhat the same way present-day archaeologists view primitive technological developments of past civilizations.

Future archaeologists probably won't find our burial grounds and kitchen middens of much value. Rather, it is likely that we shall be known to our very distant descendants mainly through a haphazard sample of "hard" technology preserved in the environments least used by man—the depths of the oceans and the depths of space.

The space vehicles "most likely to survive" are those that have been placed in independent orbits about the Sun or in high orbit about other planets or set down softly on the surfaces of the Moon, or perhaps even Mercury. Thus the present inhabitants of Earth may become identified in the minds of later millennia as "The First Planetary Exploration Culture" and known to them primarily in terms of those space vehicles recovered at that time and returned to the Earth for analysis.

In any case, the archaeologist's view of space is a useful one even now to help develop perspective on space activities regardless of the actual scenario of the future. Such an approach easily distinguishes U.S. from USSR space technology. Even with unmanned vehicles, the Soviets characteristically preserve the terrestrial atmospheric environment within a heavy-pressure vessel. Thus there is a minimum requirement for adaptation of internal equipment. The United States has, instead, adapted

3

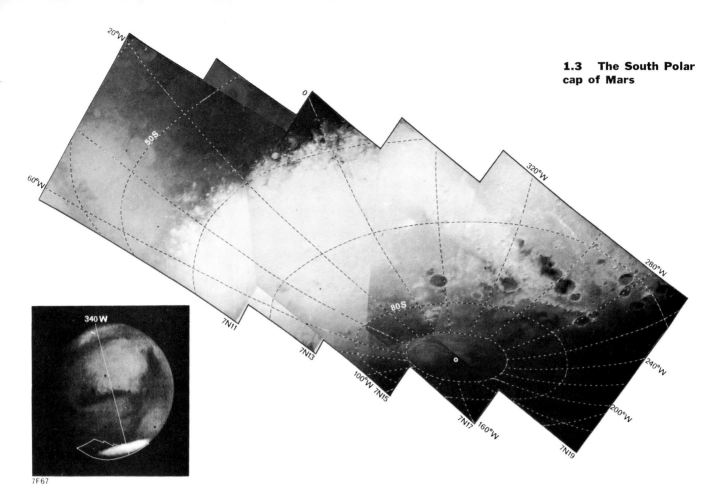

A mosaic of five television frames transmitted from the U.S. spacecraft Mariner 7 as it flew by Mars on August 5, 1969. Coordinate grid lines and the location of the South Pole are indicated also. The cap is composed of solid $CO_2$, dry ice, not water frost as was commonly believed before the advent of unmanned space probes. Large craters are the principal landform visible in this mosaic but higher resolution pictures also reveal unique surface features not recognized elsewhere. No clouds have been confidently identified in the mosaic although some hazes are suggested, for example in the lower left hand corner.

miniaturized electronic and other components to operate in the vacuum environment of space itself. American space vehicles characteristically employ more ambitious and more complex technology; yet, the United States has also built the role of a human pilot irretrievably into its manned space systems. The Soviets, despite a smaller industrial background in automation, treat the human more as a passenger, relying primarily on ground-controlled automated systems. Valentina Tereskova, the first woman in space, could not even fly an airplane! Thus the comparative anatomy of U.S. and Soviet space vehicles reveals the technological style of their builders—our hypothetical archaeologist of the future might even identify two separate "phases" within the "First Planetary Exploration Culture."

However, comparison of the U.S. and Soviet space efforts can reveal far more than style. It provides an insight into the objectives of those societies. The quality and priority of scientific instrumentation aboard a planetary spacecraft reflect the relative priority in that society of genuine cultural achievement as compared to the attainment of chauvinistic technological firsts. The relative efforts in "civilian" applications such as civil communication and weather satellites, as compared to clearly military applications, reflect the relative priorities of a society for improving the material condition of its people versus creating military power for external use. On a larger scale, the relative effort in space exploration (however chauvinistic)

4

compared to total military and civilian exploitation of space must measure to some extent the imagination and optimism of a nation's people.

Thus space is in some ways a mirror, reflecting the style, motivations, and even aspirations of the terrestrial societies which venture into it by the technologies, objectives, and priorities displayed there. The events in space of the last decade must be viewed in the context of the rivalry between the U.S. and the USSR—indeed this rivalry clearly has provided most of the priority for the development of space technology—but with very interesting national differences in approach, scale, and emphasis. During the coming decades more countries will display their individual national signatures in the mirror of space.

Furthermore, space photography must itself be viewed in terms of the anatomy of space, of technological style, of relative use in exploration versus exploitation, and of its use in manned and in unmanned systems. The number of significant space-faring nations will surely expand. The emphasis on obtaining practical military and civilian benefits versus further exploration will wax and wane as national destinies work themselves out. Furthermore, the relative importance of man in space will reflect the extent of continuing public support for and involvement in this exciting but expensive endeavor. Yet through it all, space photography will be not only a principal tool of space activity, but also the purveyor of the view from space to mankind and thus the harbinger of change in man's view of himself.

## 1.3 TECHNICAL ASPECTS OF SPACE PHOTOGRAPHY

This book, written chiefly for the general reader, is concerned largely with the results, significance, and potential of space photography. The more technical details appropriate to the interests of specialists are discussed in the four appendices following Chapter 7. In a similar vein, we have usually followed the practice of *Scientific American* in omitting detailed references and instead supplying selected references at the end of each chapter. We have further tried to make each chapter relatively independent so that each can stand by itself.

There are, however, a few basic technical concepts of space photography which pervade the entire book. We discuss these here briefly in the hope that, combined with the general technical background accumulated by so many in our society, they will prove adequate for most readers.

Practically everyone in the United States owns, or at least has used, a film camera. And certainly no reader of this book is unfamiliar with television display and viewing. Thus most readers already possess some technical background appropriate for this book. Indeed, the fact that amateur photography is so widespread has, on occasion, hindered the development of professional approaches to space photography.

5

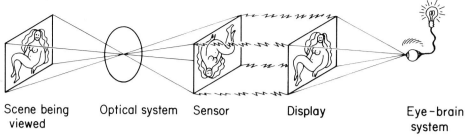

Scene being viewed    Optical system    Sensor    Display    Eye-brain system

The transfer of information about a scene to the eye-brain combination is illustrated schematically. In film photography at home or by astronauts, the sensor is the negative and the positive print is the display. With unmanned probes, the film negative may be readout onboard, or a television image may be sensed and radioed to earth where the positive print display is usually prepared. In all cases, photography is a communication process.

Since *everyone* can be a photographer, surely *anyone* can work out the details of taking pictures from satellites! However, space photography is more sophisticated than everyone's home camera experience might indicate. It must be recognized that a photograph is not an end in itself. A photograph is an intermediate recording and filtering of the brightness distribution in the scene viewed by the camera. When that representation of the visual scene is, in turn, viewed and interpreted by the human eye-brain combination, then and only then does a photograph have meaning. A photograph thus can be regarded as nothing more nor less than a link in a communication system (see Figure 1.4). This is a particularly appropriate basis from which to consider space photography because the intrinsic difficulty and expense of the return of photographic information from space has demanded a most careful review of what signals really need to be transmitted by the communication system. What is it we wish to know about the visual scene as viewed from space? How can that information be most effectively recorded and returned to the ground? How can the view then be reconstructed to relay to the human observer all the information actually transmitted? These are the basic questions that have confronted those who have photographed from space the Earth, the Moon, and the planets. In the course of this book, we will illustrate many of the ingenious technical schemes that have been developed and also examine the results

of these efforts, that is, the photographs taken from space.

To discuss this communication process, we need a unit of measure. Somehow, we must be able to count how many tiny pieces make up any particular picture. More specifically, we need to know how wide a communication bandwidth and how long a time will be required to transmit a picture without appreciable degradation. Appendix A presents a formal theory for this process. It will suffice for our needs here to state that the *bit* is the unit of measure used by specialists in communications. As applied to pictures, one can imagine the image to be made up of a multitude of tiny black or white dots in a regularly spaced two-dimensional array much like the common newspaper photoprint. The tonal level of a particular part of a picture viewed by the eye is governed by what fraction of those dots are black and what fraction white. Indeed, the semitone illustrations of this book are created in an analogous manner. Now if we wished to transmit such a picture by radio to a distant point, we would need merely to transmit a series of zeros and ones (perhaps as "off" and "on" signals) to represent the black and white dots. Then the person receiving the transmission could recreate the picture by converting each zero (or "off") to a black dot and each one (or "on") to a white one, and so on. The term *bit* could be applied to each zero-or-one piece of the message. There is more to the question than implied by this oversimplified example, as is discussed in

| Table 1.1 Photographic exploration of Mars (all data approximate) | Total data return (bits) | Best surface resolution (km) | Total cost[a] ($) |
|---|---|---|---|
| All ground-based observations (telescopic) | $7 \times 10^{7}$[b] | 100 | ? |
| Mariner 4 (1965) (flyby) | $3.5 \times 10^{6}$ | 3 | $1.25 \times 10^{8}$ |
| Mariners 6 & 7 (1969) (flybys) | $5 \times 10^{8}$ | 0.3 | $1.5 \times 10^{8}$ |
| Mariner Mars '71[c] (orbiters) | $5 \times 10^{10}$ | 0.06 | $1.5 \times 10^{8}$ |

[a]Assuming entire mission cost is to be borne by photography even though other instruments are also carried.
[b]Estimated from Murray (1965)
[c]Projected data based on current plans.

Appendix A. However, we will speak here in terms of the total number of bits which constitute a single frame, or, for example, have been received from a single Mars mission.

In Table 1.1 this tabulation of total photographic data in terms of bits is illustrated. There, and in Figure 1.5, the enormous increase in photographic knowledge of Mars is depicted, along with the extraordinary reduction in unit cost of photographic data. Thus it is apparent that, in terms of our information about Mars, we are in the midst of a revolution in which the rate of acquisition of significant new data is straining the ability of scientists to really understand their meaning. We shall illustrate repeatedly in the following chapters how discovery has come about again and again through sheer force of overwhelming data despite the occasional difficulty for all-too-human scientists to recognize new and unexpected patterns.

Before recounting these events of planetary exploration (Chapters 4, 5, and 6) it will be desirable to examine the role of space photography in Earth applications (Chapter 2) and lunar exploration (Chapter 3). Finally, in Chapter 7, we will attempt to discuss the future of space photography in the context of the future of space endeavors generally.

### REFERENCES

Murray, Bruce, "A Martian Horror Story; Requirements for the Photographic Exploration of Mars," *Advances in Astronautical Sciences, 19,* 1153–73, 1965.

### 1.5 Mars information revolution

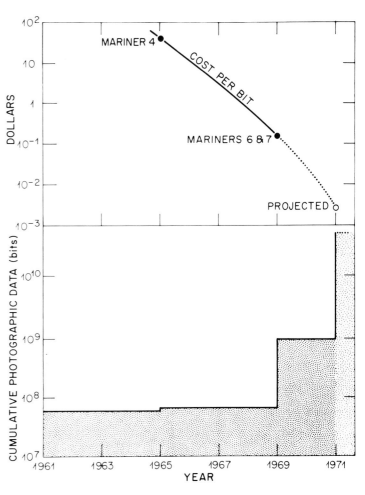

The extraordinary growth of photographic information about Mars is illustrated in the lower illustration and the equally rapid decrease in unit cost of that photography is displayed in the upper. Data and references are in Table 1.1.

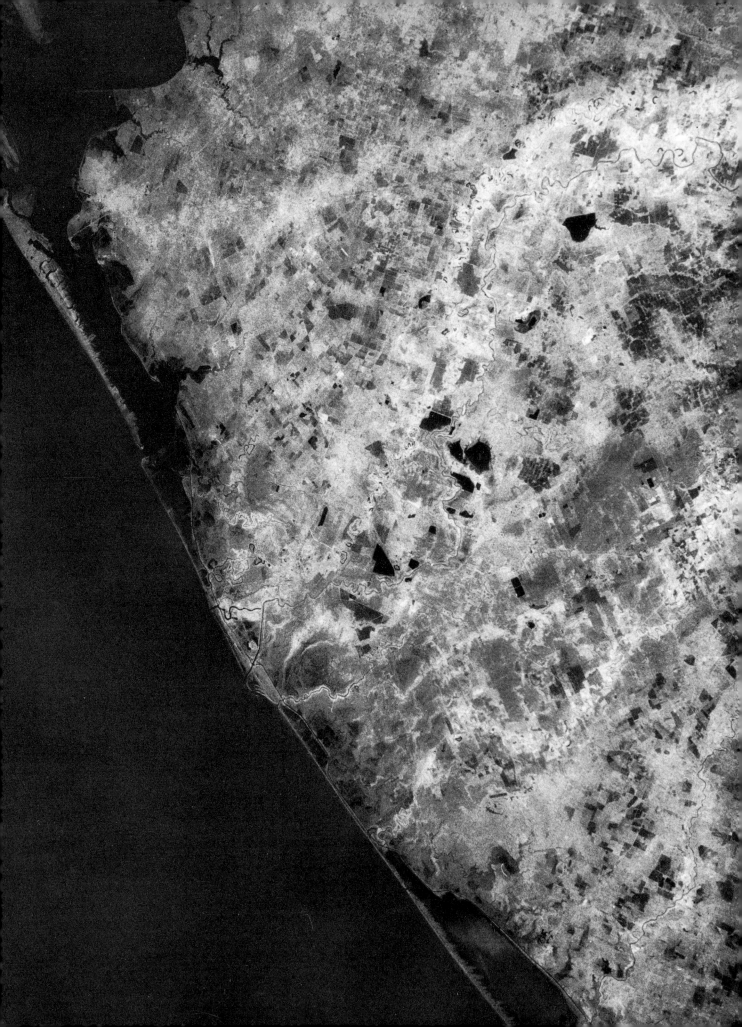

# CHAPTER 2
# Inspection of Earth from orbit

## 2.1 THE CLOUDY EARTH

The meteorological satellite program started early in the space age and has developed into one of the most valuable space programs in its service to earthlings. Currently this program is playing a major role in collecting data on a global scale and delivering it to data centers for analysis and distribution. With improved sources of data, meteorologists will increase their knowledge of the physics and circulation of the atmosphere, permitting ever more accurate long-term worldwide forecasts to be made. Improved methods for weather modification also are being studied and will be increasingly used in the future, so it will be important to monitor and predict the global effects of regional weather changes.

The role that satellites might play in collecting meteorological data was first examined by careful study of cloud pictures obtained by cameras mounted on v-2 and Aerobee rockets fired at White Sands after the end of World War II. This study by Greenfield and Kellogg, both then at Rand, investigated the possibility of obtaining useful synoptic meteorological data from satellites and showed how such data could be utilized. In this connection they examined useful orbits as well as the resolution requirements needed for useful cloud pictures and suggested that the United States should launch a meteorological satellite when the necessary technology was available. The Defense Department did start such a program,

which was transferred to NASA, and led to the development and successful launching of the TIROS I satellite on April 1, 1960.

The TIROS program was experimental and supplied the experience and data upon which to base future designs. Most important, the TIROS satellites provided the incentive to develop techniques and procedures for integrating satellite data into the existing worldwide meteorological data net—a task not yet completed. Nine TIROS satellites were launched before the Department of Commerce's ESSA (Environmental Science Services Administration) system became operational in 1966 (see Table 2.1). All of the TIROS satellites were spin stabilized with the spin axis in the orbital plane. They contained two vidicon television cameras, one wide angle and one narrow angle, with the optical axes parallel to the spin axis. With this arrangement, the cameras viewed the earth on less than half of each orbit. Images were tape recorded for subsequent transmission to earth. As new models were built, improvements were incorporated. These included Automatic Picture Transmission, which permits direct readout by a ground station of a picture containing the surrounding region, scanning radiometers, and finally, change of the spin axis from one in the orbital plane to one normal to the orbital plane so that the satellite, when in orbit, revolved like a wheel. With this attitude change, the television cameras were mounted normal to the spin axis so that they would see the earth at nadir on each revolution.

9

| Table 2.1 U.S. meteorological satellites | Name | Launch date | Weight (kg) | Period (min) | Perigee (km) | Apogee (km) | Incl. (deg) |
|---|---|---|---|---|---|---|---|
| | TIROS 1 | Apr. 1, 1960 | 119 | 99.2 | 692 | 753 | 48.3 |
| | TIROS 2 | Nov. 23, 1960 | 126 | 98.3 | 623 | 727 | 48.5 |
| | TIROS 3 | July 12, 1961 | 129 | 100.4 | 742 | 814 | 47.8 |
| | TIROS 4 | Feb. 8, 1962 | 130 | 100.4 | 710 | 845 | 48.3 |
| | TIROS 5 | June 19, 1962 | 130 | 100.5 | 591 | 972 | 58.1 |
| | TIROS 6 | Sept. 18, 1962 | 127 | 98.7 | 681 | 715 | 58.2 |
| | TIROS 7 | June 19, 1963 | 135 | 97.4 | 620 | 645 | 58.2 |
| | TIROS 8 | Dec. 21, 1963 | 120 | 99.3 | 692 | 761 | 58.5 |
| | NIMBUS 1 | Aug. 28, 1964 | 376 | 98.3 | 423 | 932 | 98.6 |
| | TIROS 9 | Jan. 22, 1965 | 138 | 119.2 | 700 | 2578 | 96.4 |
| | ESSA 1 | Feb. 3, 1966 | 138 | 100.2 | 695 | 838 | 97.9 |
| | ESSA 2 | Feb. 28, 1966 | 132 | 113.6 | 1357 | 1424 | 101.0 |
| | NIMBUS 2 | May 15, 1966 | 414 | 108.1 | 1101 | 1181 | 100.3 |

Complete orbital coverage was obtained. The ESSA operational satellites also use this stabilization mode and have incorporated a new high-resolution one-inch vidicon tube.

The NIMBUS satellite was developed during the same period as TIROS; it was larger, more complex, and incorporated earth-oriented stabilization. NIMBUS continues to be used as a test bed for advanced instrumentation.

The early TIROS satellites were launched in orbits with inclinations of about 48° and an altitude of about 700 km. To obtain worldwide coverage, the later TIROS and ESSA satellites were launched on slightly retrograde near-polar orbits at a 1200 km altitude. This retrograde orbit will precess, due to the pull of the Earth's equatorial bulge, at just the proper rate so that the Sun is at the same elevation each revolution throughout the year. This slightly retrograde orbit (97° to 101° inclination depending upon altitude) is important to all picture-taking satellites for maintaining proper lighting over long lives (see Fig. 2.1).

ESSA operates two readout stations, called Command and Data Acquisition Centers, located at Fairbanks, Alaska, and Wallops Island, Virginia, which relay the pictures to the National Environmental Satellite Center at Suitland, Maryland. With the help of a computer, cloud mosaics are drawn which show the global distribution of clouds. These are then transmitted by radio-facsimile from the World Meteorological Center in Washington to the corresponding center in Moscow.

ESSA has launched satellites with two different types of payload. In one type, the tape recorder stores television pictures taken in daytime and infrared radiometer pictures taken at night, which are then read out at a Command and Data Acquisition Center. The other configuration contains the APT (Automatic Picture Transmission) system, which transmits cloud pictures continuously and may be received by any ground station, including ships at sea. Normally a station can receive three pictures per orbit for three orbits per day, permitting cloud coverage for a radius of over 3000 km from the station to be observed. The APT ground station contains an antenna, receiver, and facsimile recorder; there are several hundred such installations throughout the world using these pictures daily for local forecasting.

The ATS 1 and ATS 3 synchronous satellites

| Name | Launch date | Weight (kg) | Period (min) | Perigee (km) | Apogee (km) | Incl. (deg) |
|------|-------------|-------------|--------------|--------------|-------------|-------------|
| ESSA 3 | Oct. 2, 1966 | 145 | 114.5 | 1384 | 1485 | 101.0 |
| ATS 1 | Dec. 6, 1966 | 352 | 660 | 35,850 | 36,885 | 0.2 |
| ESSA 4 | Jan. 26, 1967 | 132 | 113.4 | 1323 | 1439 | 102.0 |
| ESSA 5 | Apr. 20, 1967 | 145 | 113.5 | 1352 | 1421 | 101.9 |
| ATS 3 | Nov. 5, 1967 | 365 | 1436.4 | 35,772 | 35,813 | 0.4 |
| ESSA 6 | Nov. 10, 1967 | 133 | 114.8 | 1410 | 1489 | 102.1 |
| ESSA 7 | Aug. 16, 1968 | 145 | 114.9 | 1431 | 1469 | 101.7 |
| ESSA 8 | Dec. 15, 1968 | 132 | 114.6 | 1416 | 1464 | 101.8 |
| ESSA 9 | Feb. 26, 1969 | 145 | 115 | 1421 | 1518 | 102 |
| NIMBUS 3 | Apr. 14, 1969 | 576 | 107.3 | 1070 | 1131 | 99.9 |
| ITOS 1 | Jan. 23, 1970 | 373 | 115 | 1436 | 1482 | 101.9 |
| NIMBUS 4 | Apr. 8, 1970 | 621 | 107.1 | 1093 | 1102 | 99.8 |
| NOAA 1 | Dec. 11, 1970 | 309 | 114.8 | 1429 | 1472 | 101.9 |

contained line-scan cameras which completed a picture of the Earth in about 20 minutes. The satellites are spin stabilized and the picture is read out line by line as the lens sweeps across the Earth during each revolution. ATS 3 contains filters so that color pictures can be assembled. These pictures are very useful because the satellite sits over the same spot on the equator and records cloud movements by successive frames, thus providing a "time-lapse" view of atmospheric phenomena, including major storm systems, as revealed by the changing cloud formations over about one-third of the Earth's surface (see Figs. 2.2 and 2.3). The technology represented by these experimental satellites is very impressive and

### 2.1  Effect of seasons on solar illumination

Top—90° inclination orbit is fixed in inertial space so the space craft can observe the bright sunlit Earth only part of the year. Bottom—97°–101° inclination orbit can be chosen to regress at the proper rate so the spacecraft can observe the bright sunlit Earth throughout the year.

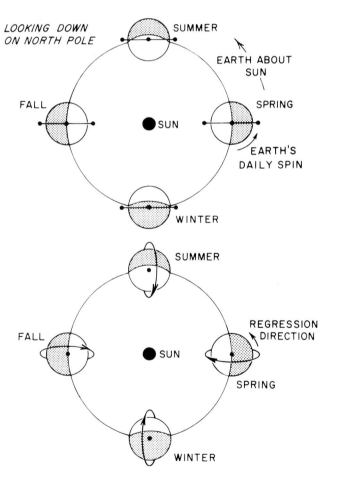

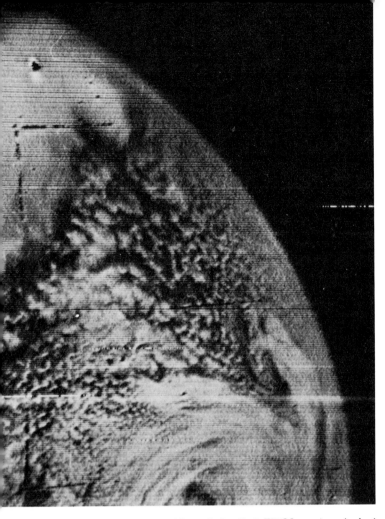

Since the launching of the first TIROS, meteorological satellites have been used to track and monitor severe storms, thus alerting populations to this source of potential danger.

no doubt points the way for future meteorological data acquisition.

The Soviet weather satellite program started later than the U.S. program with the launching of Cosmos 122 on June 25, 1966. This experimental satellite was launched on a 65° inclination and was a test of the spacecraft design. The bus, which contains the sensors, is a pressurized cylinder with its axis stabilized normal to the earth. The large externally mounted solar panels move relative to the bus so as always to face the Sun. Looking down from the end of the cylindrical spacecraft are two vidicon cameras which take pictures of the clouds in daylight, and an infrared scanner

for obtaining cloud pictures at night. Later spacecraft of this type (Cosmos' 144, 156, 184, 206, 226) were placed in near-polar orbit (81° inclination) so as to give coverage of the entire earth. This weather satellite system is now operational and the spacecraft are given the designation Meteor (see Table 2.2).

The early Molniya 1 spacecraft, of the first Soviet communication satellite system, contained vidicon cameras which transmitted full view pictures of the earth from deep space. These might have been useful for studies of worldwide circulation (as were the ATS satellites); however, the resolution was rather poor for this purpose. No cameras are reported on the more recent Molniya 1 spacecraft, so it is assumed that this phase of the project has been abandoned. Some of the Soviet lunar exploratory spacecraft took high resolution full-view pictures of the Earth after leaving the vicinity of the Moon and upon approaching the Earth (see Fig. 2.4).

Weather, knowing no national boundaries, has long been a subject for international cooperation. Satellites are destined to play an increasing role in the two major programs of the World Meteorological Organization—the continuing and evolutionary "World Weather Watch," and the massive "Global Atmospheric Research Program" (GARP) scheduled for the middle and late 1970s. Synoptic measurements of atmospheric parameters are made and reported by the many meteorological stations throughout the world. However, there are many

One of a large number of high quality full-Earth views acquired by a line-scan camera aboard the Advanced Technology Satellite #1. This particular view was acquired at about 6:45 A.M. Pacific Standard Time on April 3, 1968. The satellite is geosynchronous. North and South America comprise the central part of the picture.

A black and white copy of one of the color photographs acquired by the spacecraft Zond 7 on its outward journey to the Moon. Zond 7, like others in the series, flew around the Moon and back to the Earth where an instrument capsule was recovered directly on the surface. Included in the instrument package were the exposed negatives of this and other pictures of the Earth and Moon. The above frame was taken on August 8, 1969 at a distance of about 70,000 km. The Sinai Peninsula, also shown in Fig. 1.1, is easily recognized in the center of the frame.

areas, principally oceans and regions of the southern hemisphere, where data are particularly scarce. Thus the satellite cloud pictures were the first to monitor equally the entire planet's surface. One direction of current meteorological research is toward the use of satellites to store and read out to the data centers telemetry received from buoys anchored at sea, unmanned observatories in inaccessible land areas, and balloon platforms floating in the atmosphere.

## 2.2 PROFIT FROM PICTURES

Meteorological satellites have been in operation for more than a decade, and the number of pictures taken of the Earth runs into the millions. These systems are designed to take cloud pictures over large areas. Surface resolution (measured in kilometers) is sacrificed to

| Table 2.2 Soviet meteorological satellites | | | | | |
|---|---|---|---|---|---|
| Name | Launch date | Period (min) | Perigee (km) | Apogee (km) | Incl. (deg) |
| Cosmos 44 | Aug. 28, 1964 | 99.5 | 618 | 859 | 65 |
| Cosmos 58 | Feb. 26, 1965 | 96.8 | 581 | 658 | 65 |
| Cosmos 100 | Dec. 17, 1965 | 97.7 | 650 | 650 | 65 |
| Cosmos 118 | May 11, 1966 | 97.1 | 641 | 641 | 65 |
| Cosmos 122 | June 25, 1966 | 97.1 | 624 | 624 | 65 |
| Cosmos 144 | Feb. 28, 1967 | 96.9 | 624 | 624 | 81.2 |
| Cosmos 156 | Apr. 27, 1967 | 97 | 629 | 629 | 81.2 |
| Cosmos 184 | Oct. 25, 1967 | 97.1 | 636 | 636 | 81.2 |
| Cosmos 206 | Mar. 14, 1968 | 97 | 629 | 629 | 81 |

**2.5 Ice floes in the Gulf of St. Lawrence**

As an aid to navigation, the meteorological satellites are being used in some parts of the world to monitor the breaking up of ice shelves. This picture was taken by TIROS II.

obtain this large cloud coverage, and filters are selected and exposure controlled for good quality. However, since the beginning of these programs, cloud-free pictures of the surface of the Earth have been received which proved to be of considerable interest and led to studies of other potential uses of the pictures. Geographers, foresters, map makers, and geologists all expressed interest in the pictures, which introduced the view-from-space to a large family of scientists. It was clear, however, that better quality was needed before this new data would have a significant impact on their scientific disciplines.

There is one area, nevertheless, in which weather satellite pictures are now used for surface observation—monitoring the breaking up of ice in navigable rivers and in the polar regions (see Fig. 2.5). Ships operating in

these areas receive the ice reports from data centers, from a nearby APT readout station, or from their own APT station onboard. As might be expected, in some cases it is difficult to distinguish ice from clouds; however, IR sensors are sometimes helpful in resolving the ambiguity.

Cameras were also carried on most of the manned space missions, and selective color pictures of the Earth were taken. These photographs had better resolution than the meteorological satellite pictures and have potential uses far beyond their original ones.

| Name | Launch date | Period (min) | Perigee (km) | Apogee (km) | Incl. (deg) |
|------|-------------|--------------|--------------|-------------|-------------|
| Cosmos 226 | June 12, 1968 | 96.9 | 603 | 650 | 81.2 |
| Meteor 1 | Mar. 26, 1969 | 97.9 | 644 | 713 | 81.2 |
| Meteor 2 | Oct. 6, 1969 | 97.7 | 631 | 690 | 81.2 |
| Meteor 3 | Mar. 17, 1970 | 96.4 | 555 | 643 | 81.2 |
| Meteor 4 | Apr. 28, 1970 | 98 | 625 | 709 | 81.2 |
| Meteor 5 | June 23, 1970 | 102 | 863 | 906 | 81.2 |
| Meteor 6 | Oct. 15, 1970 | 97.5 | 633 | 674 | 81.2 |
| Meteor 7 | Jan. 20, 1971 | 97.6 | 630 | 679 | 81.2 |
| Meteor 8 | Apr. 17, 1971 | 97.1 | 610 | 633 | 81.2 |

Further studies have developed many interesting ideas for applications of space-taken pictures. However, no scientific application other than meteorology has, to date, been sufficiently important to afford its own picture-taking satellite system. To proceed with the research and to test and collect information for many potential users, NASA is developing the Earth Resources Technology Satellite. Because of the large number and diverse needs of scientists doing research on possible uses of satellite pictures, it will not be possible to satisfy everyone with regard to sensors or operational characteristics. Other space programs, such as Nimbus, can help collect data for these studies.

The National Academy of Science has sponsored extensive studies of potential uses of data from earth resources satellites; the principal areas of interest which have been identified are cartography, agriculture and forestry, oceanography, geology, and hydrology. Photographs taken from aircraft have been used operationally for a long time in all of these areas so there is considerable experience in the extraction of useful information from pictures. Recent interest in the possible availability of satellite pictures has stimulated research on the development of new and improved techniques for photointerpretation.

Photography from satellites can be useful in the preparation and updating of maps and charts. Moreover, existing organizations could immediately incorporate these materials into their files; compilers and interpreters do not care whether the pictures come from aircraft or satellite.

Agricultural studies are underway to develop techniques to perform soil and crop surveys, to detect diseased plants, to make forest surveys, to detect forest fires, to examine irrigation patterns, to observe changes in land use, and more. Until satellite systems become operational, many of these tasks could be performed by using aircraft to supply the input data. Large facilities and many interpreters will be needed to make these studies, and the market is uncertain. A few years ago the OAS sponsored a number of land-use surveys in South American countries (using aircraft-collected materials). It would be interesting to learn the value of such projects to the areas involved.

Monitoring the oceans at visual, IR, and microwave frequencies could yield data useful for ship routing such as sea state, locations of ice floes, current patterns, and surface temperature information. This information must be collected in a timely fashion, and a staff must be continuously examining the many images.

With regard to geology, a large industry exists which is concerned with petroleum and mineral exploration. These industries have developed remote sensing techniques for locating interesting local areas for surface examination; aerial photographs are commonly a primary part of this process. Satellite pictures conceivably also might be used for regional studies if the quality were adequate and the price competitive with aerial pictures. On the other hand, satellite photography probably will not be so important to this industry as was the introduction of satellites to the communication industry.

Hydrology studies deal with monitoring surface water in lakes and rivers and examining snow and ice deposits to aid in making estimates of future water levels. Such programs are important to hydroelectric power stations, irrigation systems, and flood-control measures. A photointerpretation center would be required to support this work, but it would seem that a ready market would exist for the information. The market could be tested easily by supporting the interpretation of aircraft-acquired photographs over a specific region for a year or two.

Satellites of a type to support the described earth resources program take many pictures covering large areas, often repetitively at fairly short intervals (measured in days, not minutes). A large organization with many interpreters and library facilities will be needed to study and report on the various subjects of interest. Such an organization should be permitted to grow at a rate commensurate with the value of its products. It is likely that the ground organization will be more expensive than the spaceborne component and should be designed with care. The library should contain all sources of relevant information to support the studies being made.

Meteorologists have had some experience with large and rapid processing of pictorial data and have developed standardized reporting displays. Another useful device used in the meteorological program is the APT system by which a ground station anywhere in the world can read out a television picture of the surrounding ground area. The elegance of this system is that pictures can be obtained directly by a customer without going through a central data center. A similar system might be developed for monitoring select phenomena which are of primarily local interest, like industrial pollution of rivers and bays or ice conditions in the Arctic or Antarctic waters. Since such pictures must have higher resolution than weather pictures, the ground coverage would be much smaller. Pointing would be needed to obtain pictures of the desired region, and could be programmed by arrangements with the satellite control center.

Pictures of Earth from satellites undoubtedly will serve in many ways; the remaining questions concern the rate of introduction of their use. The cost of producing useful information from pictures must be carefully examined to be sure that it is compatible with the value of the data.

## 2.3 PHOTOGRAPHY AND ARMS CONTROL INSPECTION PROPOSALS

Man has long dreamed of a world without war, and governments have periodically entered negotiations hoping to find a formula, a

17

means to resolve differences without war, a means to achieve security without arms. Since World War II the dangers of uncontrolled proliferation of nuclear weapons and unlimited stockpiles have been pointed out many times. As a consequence, various groups and governments have supported studies, meetings, and negotiations aimed at finding measures which could be implemented to decrease the threat of nuclear holocaust. It is not the intent here to argue the advantages and disadvantages of the various ideas, suggestions, and proposals which have been made over the past years. Rather, it is the intent to discuss the characteristics and capabilities of observation (picture-taking) satellites as a means to inspect compliance with arms agreements. It is natural, in the space age, to look to Man's newest technology for help in resolving some of the most difficult political problems of the age.

Since 1961 the focal point for negotiation and coordination of ideas related to arms control within the United States has been the Arms Control and Disarmament Agency. Negotiations have been taking place within the United Nations since its founding; currently they are being considered by the Eighteen Nation Disarmament Committee which meets in Geneva. Extensive negotiations have resulted in treaties signed by groups of nations; these are the Limited Test-Ban Treaty, Outer-Space Treaty, Latin American Nuclear Free Zone Treaty, and Nonproliferation of Nuclear

Weapons Treaty. Since the beginning, a common stumbling block to agreements has been the need for verification in an inoffensive and unobtrusive manner. To get a feel for the role which inspection satellites might play in the future, it is useful to review some of the past suggestions which depended upon the use of aerial photography and observation satellites to verify a nation's compliance with terms of a treaty. And it is important to point out certain constraints to aircraft and satellite operations which affect the ability of these vehicles to perform specified inspections.

For photographic inspection, aircraft flying at high altitude have a tremendous advantage over satellites because they can be sent to a particular area when desired. By contrast, the earth trace of the satellite is governed by orbital mechanics, and it is difficult to deviate from that orbital path after launch. Cloud cover can interrupt photography of the ground from either vehicle; however, pictures from meteorological satellites can be used in planning the photographic sequencing. The flexibility of the aircraft operation permits more efficient use of this information than the satellite systems. (Areas near the poles which are in darkness many months of the year will not be subject to this type of inspection during those periods.)

There will generally be two types of requirements for satellite inspection: the first is for frequent looks at a particular small area of the earth to observe change; and the second is for an initial inventory of objects over a large area. The first mission can be accomplished by placing the satellite in an orbit where its period is an integral division of 24 hours so the earth traces repeat each day. For instance, selected periods might be $24 \times 60/16 = 90$ minutes, $24 \times 60/15 = 96$ minutes, $24 \times 60/14 = 102.857$ minutes, and so on. Then the inspection area can be viewed at least once a day in daylight (and perhaps more often, depending upon latitude). If the area is near a polar region, a high-inclination orbit will permit two, three, or more looks per day. If the inspection area is in the equatorial region, a low-inclination orbit will permit two, three, or more looks per day. The most difficult areas to inspect lie in the middle latitudes where an optimum low-altitude orbit will permit only one daylight pass per day.

The selection of orbits for the second type of mission, the large-area survey, is also very dependent upon the latitude of the region to be inspected. In the polar or equatorial regions the survey can be quite efficient if lighting conditions and cloud cover permit. However, if the inspection area covers a large spread in latitude, it is not possible to obtain a very efficient orbit. This can be seen by considering an example. Suppose the area extends from the equator to high northern latitudes. Then the orbit must be near-polar so that the northern regions can be viewed and the orbit will cross the equator every 2500 km. If the camera has 90° photographic coverage, the swath width from 200 km altitude is about 400 km on the ground. Thus, for full coverage at the equator more than six days of photography will be needed. In practice, extended operations will be necessary to obtain good pictures of a large area because of cloud cover.

The most famous of the occasional proposals for aerial or space inspection is the "Open Skies" proposal of President Eisenhower in 1955. The suggestion was to use unlimited aerial photography to verify an exchange of military "blueprints" between the Soviet Union and the United States. This suggestion, made at the July 21 Geneva Conference of heads of government, was turned down by Premier Bulganin in his address to the Supreme Soviet August 4. An outline plan for the implementation of the "Open Skies" proposal was submitted to the Disarmament Subcommittee of the United Nations on August 30. This plan called for the exchange of all data relative to military forces and installations together with measures for verification and surveillance to provide against the possibility of surprise attack. The plan was to be accomplished by posting on-the-spot ground observers and by flying unrestricted aerial photographic reconnaissance. Although nothing substantive came of these notions, they did introduce the new idea that there was an important role for overhead reconnaissance in inspection. Also inherent in this plan was the implication that there can be positive disarmament measures only when it is possible to understand, locate,

19

and confirm the existence of all military forces.

The delegates from Canada, France, the United Kingdom, and the United States submitted a paper to the United Nations Disarmament Subcommittee on August 2, 1957, on inspection to safeguard against the possibility of surprise attack. This proposal suggested that all of the territory of the continental United States, including Alaska and all of the territory of Canada, and the USSR be opened to inspection. Further, it was proposed that, if the Soviet Union rejected this broad proposal, all territory be open for inspection north of the Arctic Circle in the Soviet Union, Canada, the United States (Alaska), Denmark (Greenland), and Norway; all the territory of Canada, the United States, and the Soviet Union west of 140 degrees West longitude, east of 160 degrees East longitude, and north of 50 degrees North latitude; all the remainder of Alaska and the Kamchatka peninsula; and the Aleutian and Kurile islands (see Fig. 2.6). The inspection presupposed the establishment of ground observation posts, mobile ground teams, and aerial inspection, all designed to safeguard against the possibility of surprise attacks.

Polish Foreign Minister Rapacki, on October 2, 1957, presented his famous proposal to the General Assembly of the United Nations for establishing a nuclear free zone in central Europe encompassing Poland, the German Democratic Republic, and the Federal Republic of Germany. The proposal was amplified in a note from Rapacki to American Ambassador Beam on February 14, 1958. Czechoslovakia was included in the nuclear free zone, and inspection was to include ground as well as aerial systems. It was pointed out that the system of control established for the denuclearized zone could provide useful experience for the realization of broader disarmament agreement.

The Surprise Attack Conference took place in Geneva late in 1958 between five Western countries on one side and five Eastern powers on the other. Each side seemed to have differing views of the purpose of the meetings, so agreement could not be reached on how to proceed or even on an agenda. However, each side independently tabled papers which contained its own ideas of procedure and proposals. The Western analyses described the military forces which could be used in a surprise attack and discussed means of inspection and control of these forces. These studies, prepared by the scientific and military experts of the United States, the United Kingdom, France, Canada, and Italy, evaluated aerial and ground inspection, various other techniques, and, for the first time, the use of satellites as parts of an overall control system. The experts of the Soviet Union, together with delegates from Poland, Czechoslovakia, Romania, and Albania, proposed that a zone extending 800 kilometers to the east and west of the line separating the NATO countries and the Warsaw treaty countries as well as

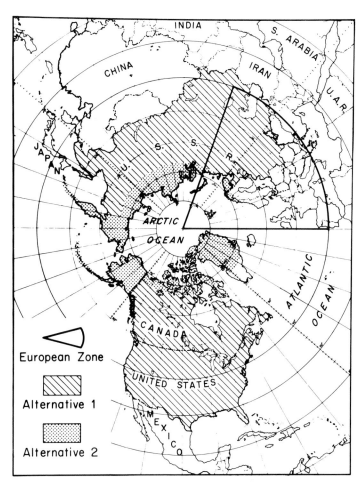

After the Soviet rejection of the "Open Skies" proposal of 1955, a modified aerial and ground inspection plan was submitted to the United Nations Disarmament Subcommittee by the governments of Canada, France, the United Kingdom, and the United States.

European Zone

Alternative 1

Alternative 2

the territory of Greece, Turkey, and Iran be established with 28 ground control posts on each side including six in the USSR and six in the United States. Aerial photography was also to be used in this zone for control. There were proposals to reduce the number of foreign troops in this area and to prohibit nuclear and rocket forces in the two parts of Germany.

Western documents discussed in detail the weaponry which would characterize forces involved in a surprise attack, and the characteristics of aerial, satellite, underwater, and ground sensor techniques which would constitute the control system. Studies were presented which evaluated the capabilities of this control system to inspect surprise attack forces. The communication requirements and data processing center needed to continually evaluate the surveillance information were also given.

Photographic inspection techniques related to aerial and satellite vehicles were divided into three categories of ground resolution considered necessary to permit identification of objects of different sizes. These were:

I.   75 to 100 ft (satellite altitude)
II.  5 to 10 ft (high altitude aircraft)
III. 1 to 2 ft (low-altitude aircraft)

Examples of operational conditions for vertical photography at a film resolution of 15 to 25 optical lines/mm, which would yield pictures corresponding to these categories, are:

| Category | Altitude | Camera lens focal length |
|---|---|---|
| I | 150 mi | 36 in. |
| II | 50,000 ft | 36 in. |
| III | 5,000 ft | 18 in. |

(From Western Experts Document presented to Conference November 19, 1958)

The analysis of capabilities and limitations of such photography to perform the search, identification, and inspection necessary for

21

control was carefully presented. For many control functions, ground teams would be required.*

On March 15, 1962, the Soviet Union presented to the Eighteen Nation Disarmament Committee a draft treaty calling for general and complete disarmament. This proposal carefully described the gradual reduction of armament and the destruction of military facilities to take place in three stages and to be carried out under strict international control and supervised by inspectors of the International Disarmament Organization. The International Disarmament Organization was to have access at any time to any point within the territory of each party to the treaty. Also the IDO would have the right to institute a system of aerial inspection and aerial photography over the territories of the treaty states. This was the only mention of aerial inspection in the proposal treaty.

The United States, on April 18, 1962, also submitted a plan for general and complete disarmament to the Eighteen Nation Disarmament Committee. This plan called for gradual disarmament in three stages and the creation of a United Nations Peace Force to deter or suppress any threat or use of arms. An International Disarmament Organization, to be operated by the United Nations, would be charged with verifying that the terms of the treaty were being carried out. Inspectors with

unrestricted access to all places would serve this function. No mention of aerial or satellite photography for inspection was made in this proposal.

Although neither the Soviet nor U.S. proposals made any important contributions to the notion of inspection by aerial or space photography, as did the Surprise Attack Conference reports, the proposals are important because they both agreed upon the establishment of an International Disarmament Organization which would be responsible for verifying that treaty measures were being carried out. The IDO would also be in charge of the inspectors, aerial photography, and other inspection techniques. This role of the IDO within the United Nations is an interesting one even when separated from the General and Complete Disarmament Proposals. Such an organization could operate a special information center during crises as well as an inspection center monitoring terms of agreements.

In the last few years there has developed an increasing interest in monitoring certain arms limitation agreements by means of unilaterial collection techniques. In 1968 former Secretary of Defense Robert S. Mc-Namara spoke out in support of a Soviet-American freeze of offensive and defensive nuclear weapons and suggested that initially each power use observation satellites and its other sources of information for inspection without on-site verification. He did suggest,

* Appendix D contains material relevant to aerial and space photography from these reports.

however, that international inspection, including on-site inspection, would be needed at a later stage. Since that time, further proposals and discussions have led to the Strategic Arms Limitation Talks (SALT) which began in 1970 and still continue at present.

Although the use of unilaterial inspection to monitor compliance to the terms of an agreement is a way to get started, there is no mechanism to resolve controversy in the interpretation of collected data. Thus suspicion may be fostered by incomplete and partial information. The technical options in the design, testing, and operation of modern weapons are so great that flexibility in inspection will be necessary if all elements of governments are to feel confident that all parties are adhering to the terms of a treaty. Thus it might be necessary to review the inspection agreements, perhaps at regular intervals, to insure that they do adequately reflect the changing technology. Adequate inspection is the real key to promoting confidence between governments which is necessary if nations are to build a world free from the threat of nuclear war.

## 2.4 A ROLE FOR THE U.N.?

Both the United States and the Soviet Union have proposed the establishment of an International Disarmament Organization to be operated by the United Nations and which will be responsible for verifying that the terms of a treaty are being carried out. These proposals suggest that the IDO will be in charge of all inspection operations, which include ground inspectors, ground control posts, and aerial and space inspection. Since both major powers have indicated in principle that an IDO would be needed to monitor treaties, perhaps it would be sensible to establish the IDO prior to treaty negotiations. Developing the inspection techniques and capabilities *before* they are needed could promote arms limitation agreements which reflect the effectiveness of the IDO inspection methods. This could be particularly important when the inspection instrument, such as a satellite, has a long lead time.

It is obvious that the techniques of inspection to be used by the U.N. to monitor specific agreements properly depend upon the nature of the agreement. Of particular interest here are arms limitation suggestions which could depend upon orbital inspection as a primary means of verification.

One important class of agreements which could use observation satellites as a principal means of inspection deals with the prohibition of certain military weapons in a specified area. One such accord, the Latin American Nuclear Free Zone Treaty, has already been negotiated. This treaty prohibits the basing of nuclear weapons in Central or South America. No inspection is provided for in the agreement. Suggestions have also been made to set aside Africa as a nuclear-free region, but no formal action has yet taken place. This type of agreement, and also one prohibiting the basing

23

of large missiles and long-range aircraft, could be monitored by inspection satellites so long as the bases were permanent facilities. Most facilities associated with fissile material production and fabrication, most nuclear weapon storage areas, launch facilities for large missiles, and bombers at airfields would be identified. Some types of operations, such as the use of large airfields for bomber staging bases, could no doubt go undetected by an inspection satellite. Such operations, performed with special care and precautions, might deceive the satellite inspector; however, failure of the ruse because of miscalculating the satellite's capabilities would be particularly embarrassing.

Another example of the usefulness of satellites for inspection would be monitoring an arms-free zone established between belligerent neighbors. Such areas might be set up by negotiation between Israel and Egypt, India and Pakistan, and other areas of the world as a mechanism to help decrease the opportunities for incidents. Satellite surveillance of this type would be much more efficient over a desert area, such as the Middle East where the sky is usually cloud-free and the foliage is scarcer, than over an area like Southwest Asia with monsoon rains and jungle foliage.

Satellite surveillance of specific ports could monitor the coming and going of ships in support of U.N. sanctions or some agreement denying a port to specific ships. Since ships are large, easy to identify, and slow moving,

tracking and docking can be nicely monitored by an inspection satellite viewing the area once or twice a day and transmitting its pictures to a suitable ground station. Thus the port activities can be reported within hours after the pictures are taken.

In a crisis, a readout satellite (television) could also serve as an additional source of information for the Secretary General. For instance, during the Six Day War of 1967 a satellite of this type could have supplied the Secretary General with pictures of the combat area once or twice a day and helped him understand what was occurring and the extent of damage. Accurate, detailed, impartial information (pictures) might help draw realistic cease-fire lines and encourage both sides to respect them inasmuch as violations might be detected and influence U.N. actions or voting.

With regard to U.N. satellites, there is the question of access to the information (pictures) and control over it to assure that it is not put to uses for which it is not intended. These pictures would be very helpful in locating the military forces of a nation, and their availability might encourage a surprise attack on a neighbor—exactly contrary to the desired result. For this reason special procedures will have to be set up within the U.N. for handling the satellite material. Of course, within the next decade many countries will have the capability of orbiting satellites and obtaining such pictures unilaterally. It would thus seem prudent for nations to assure that their military forces were not deployed in such a manner that they would be especially vulnerable to attack if such pictures fell into hostile hands. In some circumstances, there might be agreement to release the U.N. pictures because they will have great worldwide educational value and will be useful for many nonmilitary applications (see Section 2.2).

There is a good deal of merit in having a third party, the U.N., perform these types of inspection. Under certain conditions, the pictures could detect treaty violations or perhaps the aggression of one country against its neighbor. With such data at hand, the U.N. would have evidence upon which to designate blame or to base action. In this role the evidence would be much more credible than only statements by a nation charging its neighbor with hostile action.

The inspection satellite would be a modest program when measured against the large U.S. and Soviet space efforts. Pictures could be taken with television or film camera from an earth-oriented or spin-stabilized satellite. Any of the standard camera types—frame, panoramic, or strip (see Appendix B)—could easily be adapted to this mission. The pictures could be returned to Earth by electronic readout or by physical recovery of the film. Many choices regarding sensors and operations are available. In practice, costs and necessary performance would properly narrow the options.

The Lunar Orbiters (see Chapter 3.2) con-tained excellent photographic systems consisting of camera, film, a processor, and electronic readout. In operation, the orbiters achieved or exceeded their design resolution of 76 lines/ mm at 3:1 contrast, thus returning beautiful pictures of the Moon. It might be imagined that this same camera could be put in an earth satellite. Pictures of the Earth taken with the 610-mm focal-length lens from an altitude of 200 km would have a ground resolution of about 4.3 m when transmitted to the ground stations and properly reconstructed. Another mode of operation would be to expose the film in space and return it physically to Earth for processing, perhaps in a manner similar to the method of recovering monkeys with the bio-satellite. In this case, the resolution of the film would be 120 lines/mm for the Lunar Orbiter camera system, for it was the scanner, not the film, which limited the performance to 76 lines/mm. Thus this camera, with film recovery, would produce pictures with a ground resolution of 2.7 m from 200 km altitude.

For purposes of inspection from satellites, a ground resolution of the order of 2 m would be very useful. This could be achieved with the Lunar Orbiter photographic system if the film were returned to Earth and the altitude at the time of photography reduced to 146 km. If this were undesirable for some reason (such as limiting the orbital lifetime), a new 830-mm focal-length lens could be designed for the camera so the system could yield the 2 m resolution from 200 km altitude.

LAUNCH PADS U/C    CONTROL BUNKERS U/C

**2.7 Before and after construction of missile base (Cuba 1962)**

This IRBM launch site near Guanajay, Cuba, was photographed in 1962 by aircraft flying at high altitude before construction started (left) and during construction (right). The original pictures are reproduced here at a ground resolution of two meters to simulate satellite photography. The form and layout of the earth scars are used to identify the nature and purpose of the construction. The later destruction of the base could not be inspected adequately at this resolution.

Figures 2.7 and 2.8 are pictures taken from high-altitude aircraft during the 1962 Cuban missile crisis that have been reprinted so that the resolution is degraded to 2 m. Thus they simulate pictures which could be taken from a satellite at this resolution. Figure 2.7 shows the Guanajay area before and after the start of a missile base at site No. 2. The pattern of earth scars forms a clear identifying signature for this type of soft base. Other missile-launching configurations, such as hard silos, would also be conspicuous during construction at this resolution because of the characteristic excavation scarring. Ancillary facilities, such as control bunkers, warhead storage, and maintenance areas, also tend to confirm the

**2.8 San Julian Airfield (Cuba 1962)**

This picture of the San Julian Airfield was taken by an aircraft flying at high altitude during the 1962 Cuban missile crisis and is here reproduced at a ground resolution of two meters to simulate satellite photography. The inset is enlarged and reproduced at the original resolution so it is possible to note the loss of data at the two-meter resolution. For inspection purposes the two-meter resolution would be sufficient to call attention to the assembly of crated aircraft; however, more information would be needed to determine the exact number of aircraft and the progress of the work.

nature of the activity, as does observation of any special missile transportation fixtures. In general, it should be possible to detect and locate permanent fixed missile bases during construction at 2 m resolution.

Figure 2.8 of San Julian Airfield shows how distinctly an airbase can be viewed at 2 m resolution. The runways, taxi strips, parking areas, and service buildings can be readily seen. Also, the surface-to-air defensive missile site (SA-2) surrounded by a security fence is easily identi-

fied. Some individual aircraft can be seen, but they would be difficult to classify except by size.

It might be important for the United Nations to be independent of both the United States and the Soviet Union in acquiring and operating inspection satellites. Moreover, redundant sources of launchers, spacecraft, tracking facilities, and data-handling facilities would be desirable so that U.N. operations could not be easily halted at the whim of any particular country; multiple sources of spaceborne and ground facilities should thus help. It is therefore useful to examine the non-U.S./USSR space efforts to see what nations might be capable of contributing to this inspection.

The most advanced launcher currently under development which might be used to boost inspection satellites into orbit is the Europa. This launcher is being developed by the European Launcher Development Organization (ELDO), which is jointly supported by Britain, France, West Germany, Italy, Belgium, the Netherlands, and Australia. The Europa I, the first configuration of this family, has already undergone a series of test flights from the rocket range at Woomera, Australia, and should place its first satellite in orbit during 1971 from a launch pad being built at the French rocket test center at Kourou, Guyana, South America. The Guyana Space Center is located very near the equator and will be used to launch the Europa II, which will boost the German-French Symphonie communications satellite into synchronous orbit. Thus launch facilities (pads, block-houses, and assembly buildings) for the Europa booster already exist at two ranges.

The Europa uses the British Blue Streak for a first stage, the French Coralie for a second stage, and the West German Arstris for a third stage. The experimental satellites with instrumentation are supplied by Italy, and third stage telemetry is furnished by the Netherlands. Belgium contributed the downrange ground-control instrumentation. The Blue Streak launch facility at Woomera has been adapted for multistage operations by Australia to accommodate the Europa.

ELDO's Europa booster would appear to be ideal for the inspection task; it can put a large weight, 1000 kg, in low orbit and will soon be operational. There is one problem. In April 1968 the British government announced that it was going to withdraw from ELDO when the present Europa I program ends, so the future availability of this launcher is uncertain.

France has a very active national space program. Her first satellite was launched November 26, 1965, from the French Space Center at Hammaguir, Algeria, using a Diamant A launcher. Three additional satellites were orbited before the French withdrew from the Sahara base on July 1, 1967, under terms of the Evian agreements. This led to the establishment of the new space center at Kourou, Guyana, where the new launch complex for the Diamant B was built. The first Diamant B successfully orbited the DIAL satellite on March 10, 1970.

No decisions have been made for advanced launchers to follow the Diamant B. Most of the studies are concerned with finding substitutes for the Blue Streak in the Europa II to launch the Symphonie communications satellite. With this long delay, it is clear that the only French launcher available during the early 1970s will be the Diamant B.

Great Britain was slow to initiate a national space launcher program. However, in 1966

**Table 2.3   The new space launchers**

| Country | Booster | First launch | Launch weight (1000 kg) | Launch thrust (1000 kg) | Payload low orbit (kg) |
|---|---|---|---|---|---|
| France | Diamant B | 1970 | 22.9 | 35.0 | 180 |
| Japan | L-4S | 1970 | 9.5 | 37.3 | 10.5 |
| | M-4S | 1971 | 43.6 | 85.0 | 75 |
| | N | 1975 | | | |
| Great Britain | Black Arrow | 1971 | 18.0 | 22.7 | 100 |
| ELDO | Europa 1 | 1971 | 105.0 | 136.8 | 1000 |
| Communist China | ? | 1970 | ? | ? | 173 |
| India | | 1974 | | | 40 |

the go-ahead was given to proceed with the development of the Black Arrow. The proposed design was an extension of the technology developed during the very successful Black Knight program. Progress has been satisfactory, and the first test firings took place at Woomera, Australia, in 1969. The first orbital attempt failed, but a second will take place in 1971.

Advanced versions of the Black Arrow have been studied, although none has been approved for development. This is not surprising since funds for the program have been scarce from its beginning and there is little reason to expect a change now. Only the need for a high priority satellite could speed up the current slow schedule.

The Japanese space program has proceeded with enthusiasm on a small budget for more than ten years. During this time, a family of high-altitude sounding rockets of increasing performance and complexity has been developed. These led to the design of the M-4S satellite launcher. While the propulsion stages of this booster were under development, it was decided to test the guidance and control system on a small rocket, the L-4S, which would still have the capability of orbiting a small payload. Four tests of the L-4S ended in failure. The fifth, launched February 12, 1970, successfully put its payload in orbit, and Japan became the fourth country to launch a satellite. After an initial failure, a M-4S successfully orbited a 63 kg satellite February 16, 1971.

The L-4S and the M-4S are being developed by the Institute of Space and Aeronautical Science of the University of Tokyo, which also operates the Kagoshima Space Center on Kyushu. Advanced models of the M-4S (M-4SH, M-4SS) are being planned and should be operational in the early 1970s.

The National Space Development Center of the Science and Technology Agency also has a space-launcher development program. This 3-stage launcher, N, will be based on the American Thor booster and will incorporate other U.S. technology as well. The first test launch is planned for 1975 and the initial launching of an operational application satellite is planned for 1977. The launch facilities for this booster will be at the Tanegashima Space Center on Tanegashima Island, south of Kyushu.

Communist China launched its first satellite April 24, 1970, and became the fifth nation to develop a space capability. No details regarding the launcher or the launch facility have been released; however, the space center is reported to be located at Shuang-ch'eng-tse in western Inner Mongolia

India has initiated a development program aimed at orbiting a 40 kg satellite in 1974. A new launch complex is under construction at Sriharikota Island in the Nellore district of Andhra Prodesh on the east coast of India. The firings will take place to the east over the Bay of Bengal.

Table 2.3 lists the non-U.S./USSR space

launchers expected to be operational in the early 1970s. With the exception of China's booster, it is likely that the United Nations could negotiate for the use of any of these space launchers as well as those of the United States and the Soviet Union to orbit inspection satellites. In fact, multiple sources would seem prudent to assure continuing inspection operations during a crisis.

The problem of designing and procuring spacecraft should be considerably easier than obtaining the launchers. Many companies and many nations are involved in satellite developments and would be competent to direct or participate in such a project. Satellites produced in Canada, Italy, and Britain have already been put in orbit by NASA boosters, as have some developed by the European Space Research Organization (ESRO). ESRO has ten members—Belgium, Denmark, France, Germany, Italy, the Netherlands, Spain, Switzerland, Sweden, and the United Kingdom—and supports a scientific research program involving the launching of sounding rockets and a few satellites. Thus ESRO permits companies and organizations in many countries to participate in a space program and to gain experience with the new technology. Many national space programs also involve building sounding rockets and their instrumented payloads. These programs, together with the national programs, assure broad experience in spacecraft design and flight instrumentation for space application.

In 1967 a design study for an inspection satellite concluded that a spacecraft weighing about 115 kg could perform a useful readout mission. This type of small satellite is compatible with the boost capabilities of the new space launchers—the Diamant B, the Black Arrow, and the M-4S—and so would seem to be a good first step in developing a U.N. inspection satellite system. The expected performance might be something like 5 m resolution at nadir from 200 km altitude. A larger satellite requiring the Europa launcher and using film with a recovery capsule could obtain a resolution of about 2 m from 150 km to 200 km altitude.

Technology does not stand still. Inspections which in 1958 (Surprise Attack Conference) were to be performed by aircraft flying at high altitude (resolutions 5 to 10 ft) will be possible from satellites in the '70s, and the political problem of negotiating for these aircraft overflights will have disappeared. Thus the '70s hold the dream of "Open Skies" for the United Nations to use as a tool in the search for peace and understanding among nations.

## 2.5 BILATERAL APPROACHES

Bilateral discussions between the Soviet Union and the United States have, so far, led to many of the existing arms agreements and will probably lead to more in the future. The Strategic Arms Limitation Talks (SALT) now underway could be a case in point. One of the most controversial aspects of previous discussions has centered on the issue of inspection and the needs to verify that the terms of the treaty are being adhered to. Frequently the United States has felt that inspection is required, whereas the Soviet Union has argued that inspection is espionage and not necessary. Thus many past discussions have been frustrated by this issue.

Things have not changed. The most controversial issue in future negotiations will still be inspection. And the ability to reach agreement on increasingly meaningful measures will to a large extent reflect confidence in the methods of verification available to the United States and the Soviet Union. Having survived a number of crises during the last decade, both nations recognize the stability of deterrence and the terrible consequences of thermonuclear war. However, the continuation of this stability lies in the ability of the two nations to maintain a careful balance between their respective numbers and types of weapons. Hence there is the need to monitor changes in strategic forces with regard to numbers and types of weapons, both offensive and defensive.

It is always difficult to evaluate the level of military uncertainty which can be tolerated so as not to upset a planned balance of power. When faced with uncertainty, it is common practice to design the military force for operation against the opponents' maximum potential threat. Since this option will not be available if limits are established on certain weapons, it is clear that the uncertainty regarding these forces must be small and that it is in the

interest of both nations to keep it this way. In this country, any talk of abrogating an arms control treaty because of uncertainty in the Soviet military capabilities would assuredly cause a heated debate in the Senate and throughout the nation.

It is not the intent to present here a full analysis of military planning in the face of uncertainty, nor of the relative tradeoffs between offensive and defensive weaponry. Rather, it is to assess the value of a satellite to perform programmed and cooperative inspections and to supply information for planning further and more detailed inspection.

The inspection satellite might be a starting point in the negotiating of an effective verification system. It could be operated jointly by the Soviet Union and the United States, or each country might operate its own system with an exchange of data, as is done with the meteorological satellites. This satellite-derived information could contain the elements of inspection for certain limited arms agreements such as the withholding of weapons from specific areas. But most important it would furnish basic material for arranging other types of inspection. The orbital parameters and pictures would supply geographical coordinates and detail locations of buildings and weapons so that an appropriate inspection can be negotiated. The agreement for further examination could take a number of forms, such as a low-altitude over-flight by aircraft, direct investigation by designated observers, or perhaps simply

31

Inspection by the U.S. Navy of missiles on the deck of the Soviet freighter **Labinsk** leaving Cuba. The Soviet sailors have partially removed the canvas tarpaulin revealing the missile in its skin-tight waterproof casing.

the display of a specific weapon at a parade or at an exhibition. The satellite pictures serve the function of isolating the area for discussion, supplying material to help define the nature of concern, and furnishing a data base for planning an appropriate inspection.

Inspection must be a cooperative venture, and initial negotiations should establish rules for inspection by aerial and ground teams. With an efficient satellite inspector, these teams should be needed only on a few occasions each year. What form might these inspections take? If it is simply a matter of photographic resolution, it should be possible to fly an aircraft with an observer and camera over the area and to return with pictures. On the other hand, the inspection of the interior of buildings and military installations would be difficult to arrange because the work in progress might be classified. This type of inspection gets to the heart of a very important question. Will it be possible to get agreements between governments as to which weapon systems, components, and facilities should properly be considered classified and beyond inspection? This is a very important and knotty question.

At least once, under the pressure of the Cuban missile crisis of 1962, U.S. and Soviet officials at the United Nations did face this problem when arranging the method and procedure for verifying the removal by ships of the missiles from Cuba. As was described in *The New York Times* of November 8, 1962, the inspections were performed by U.S. naval vessels and helicopters, which pulled alongside each Soviet freighter and observed visually and photographically the number of IRBMs and MRBMs on the freighter's deck. There was no boarding of the freighters. It had been agreed that the weapons themselves would not be examined since some parts, such as their guidance and control mechanisms, were secret. These unusual inspections were performed with the cooperation and even the cordiality of the Russians at sea. Figure 2.9 shows the *Labinsk* being inspected; the crew has removed the protective tarpaulin so that the missile in its waterproof skin-tight covering can be seen.

Another form of inspection might involve a request to open the doors of a missile silo so that the inspection satellite could take a picture to determine whether the silo was occupied or empty. In this case the satellite itself is the inspector so it is not necessary to arrange an aerial or ground inspection. However, surface observers would be necessary to confirm the absence of hidden missiles in buildings or mountain caves.

These examples suggest that certain ingredients are necessary in order to arrange meaningful inspections.

1) The party being inspected should welcome the opportunity to prove compliance with an agreement.

2) The existence of the objects of inspeciton must be established together with their approximate or suspected locations.

3) The rights and privileges of observers need to be defined in terms of the objects to be inspected because certain details of those objects must be kept secret from inspection.

4) It is reasonable to expect that different procedures for inspection will need to be arranged to answer specific questions.

Inspection is more of a continuing dialogue than a one-shot negotiation. It would be far better to provide a forum to discuss methods of inspection which could alleviate concerns rather than to permit suspicion to grow because of the limited information available. Partial or poor data could lead to the abrogation of a treaty whereas a question might never be raised if proper inspection were available. It would seem prudent to start discussions of inspection in support of arms control measures early during negotiation in an attempt to reach some working relationships before they are needed.

In summary, inspection from orbit with picture-taking satellites affords one way to help inform and thus stabilize deterrence by careful control of arms through agreements between undeceived governments. Thus one kind of modern technology can, in this instance, help free mankind from the tyranny of another kind. Such inspections could be carried out under the U.N., if that body were to assume a significant role in arms limitation agreements. Alternatively, bilateral arrangements between the U.S. and the Soviets are conceivable in which a cooperative inspection satellite system would provide a private forum in which to identify and resolve some sources of suspicion.

## REFERENCES

Arms Control and Disarmament Agency, *Documents on Disarmament*, published annually by the Disarmament Agency, Washington, D.C.

Davies, Merton E., "Big Potential for Small Observation Satellites," *Aeronautics and Astronautics*, June 1967.

Kenworthy, E. W., "U. S. Will Verify Missile Removal by Check at Sea," *New York Times*, November 8, 1962.

Kleiman, Robert, "McNamara Offers Missile-Curb Plan," *New York Times*, September 8, 1968.

McNamara, Robert S., *The Essence of Security*, Hodder, 1968.

National Academy of Science, *Useful Applications of Earth-Oriented Satellites, 13 Panel reports of the Summer Study on Space Applications*, Washington, D.C., 1969.

Raymond, Jack, "Navy Intercepts 5 Soviet Vessels in Missile Check, Weapons Sighted on 3 Ships Leaving Cuba—Russians Cordial and Helpful," *New York Times*, November 10, 1962.

Wilson, George, "McNamara Book Urges Arms Control Accord," *Washington Post*, August 10, 1968.

# CHAPTER 3
## Closer and closer to the moon

### 3.1 ZOOMING IN

As pointed out in Chapter 1, a discussion of space activities can be organized around three sets of themes: Soviet versus U.S. efforts, manned versus unmanned vehicles, and exploration versus exploitation. Exploitation is the chief application of earth orbital missions and the subject of the last chapter. Our subject here is the exploration of the moon and involves the two other themes listed above. From Luna 3 in 1959 to Apollo 11 in 1969, the prime motivation for lunar exploration by both the U.S. and USSR has been national rivalry and prestige. Unmanned techniques were rapidly expanded and deployed to prepare the way for the manned ventures to follow—and to transmit views of the lunar landscape to the millions on Earth who were intrigued to learn what their Moon looked like close up. Thus photography, first of the Moon's surface, and eventually of men walking on it, has played a central role in "the race to the Moon."

The focus of our interest in this chapter will be how space photography has increased the surface detail in our views of the Moon. The resolution of the best telescopic view was a kilometer; Ranger and Orbiter brought this down to a meter. Unmanned landers, and later the Apollo astronauts with hand-held cameras, carried the resolution well below a millimeter. Sophisticated instruments in terrestrial research laboratories have examined the returned samples at more than 10,000 times further magnification. Thus, in less than a decade man's view of the Moon has been magnified by a factor of $10^{11}$! Surely this will be one of the most rapid increases in resolution of an unknown surface ever. This breathtaking zooming is illustrated in Figure 3.1 that follows.

The first increase in resolution by a factor of 1000 came from the television cameras aboard the U.S. Ranger 7, 8, and 9 spacecraft in 1964 and 1965 as they fell to their destruction on the Moon's surface. The surface coverage provided by this zooming process was extremely limited, although adequate for determining the roughness of a few selected localities at a scale relevant to the Apollo landing vehicle, destined to follow in five years. Much more extensive high-resolution photography was achieved by the five U.S. Lunar Orbiter spacecraft in 1966 and 1967. In addition, the five orbiters produced exquisite, detailed pictures that were much superior to the low-resolution "first looks" of the reverse hemisphere of the Moon returned by the astounding Soviet Luna 3 flight in 1959, and then again by Zond 3 in 1965.

Unmanned landings began in 1966 with Luna 9, quickly followed by five successful U.S. Surveyor spacecraft and a second Soviet one, Luna 13. These robots transmitted views of their surroundings showing details of less than a millimeter in size, or an increase over the best preceding views from space by a factor of more than $10^3$. The hand-held cameras of U.S. astronauts on the surface have gone even

The sixteen frames that follow trace the appearance of the lunar surface from pre-space-age resolution to the exquisite detail seen microscopically on the surface of returned rock fragments. The series shown represents a magnification of about 100 billion fold from Figure 3.1 (a) to the part f of Figure 3.1 (k).

further with extensive surface coverage as well as photographic documentation of *in situ* materials before collection for return to Earth.

Our special interest in the Moon for this book, beyond the cultural significance of "Zooming In," is twofold. First, the lunar experience of the 1960s well illustrates the use of photographic film in space systems. Second, the abundant photography of the Moon's surface has developed the need for mapping and naming of features. How this use of photography proceeded with respect to the Moon may provide insight into future planetary activities.

## 3.2  FILM READOUT—A CLASSY TECHNIQUE

There are two ways to exploit the extraordinary resolution and information capacity of film for space photography. First, one can expose the film in space and then return it undeveloped to Earth, using either manned or unmanned transportation systems. Return to Earth, however, is expensive and complex.

In the second method, film can be developed and scanned by an automatic system aboard a spacecraft; as the scanning signal is transmitted to Earth, high quality pictures can then be reconstructed simultaneously on the ground. This latter approach, termed "film readout," has been used by both the U.S. and Soviets on the Moon and is the topic of this section. Interestingly enough, film readout was used for the lowest-quality photography as well as the highest. The epochal flight of Luna

3 in 1959 around the Moon and its return of man's first view of the reverse hemisphere of the Moon exploited a film-readout technique, as did the magnificent Lunar Orbiters which followed seven years later.

The Soviets launched Luna 3 on October 4, 1959. Their strategy called for viewing large unfamiliar areas at surface resolutions similar to those commonly achieved with Earth-based telescopes. To maximize the coverage, photographs were taken of the full Moon with portions of the Earth-facing hemisphere included to permit accurate positioning of the new pictures relative to known points.

The Luna 3 spacecraft was placed on a flyby trajectory passing within 7900 km of the center of the Moon. It continued on this path until it lay on the line connecting the Sun and Moon, at which time its orientation system was activated by ground command. Gyroscopic sensors stopped the spin and the solar sensors acquired the Sun. In the other direction, the lunar horizon sensors locked on to the moon and tracked it throughout the picture-taking sequence.

Luna 3 took fifteen pictures of the Moon. The space vehicle recorded the images on standard 35-mm film from a distance of 65,200 to 68,400 km on October 7. The odd-numbered frames were taken with a 200-mm focal length lens; the diameter of the Moon on the film was about 10 mm. The even-numbered frames were taken with a 500-mm focal length lens, and the diameter of the Moon on these

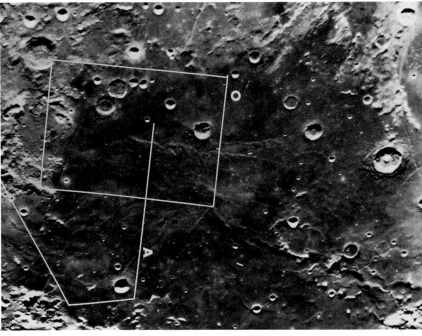

├──────────────┤
1000 km

├──────────────┤
100 km

Figure 3.1 (a). Full Moon seen from Earth. A typical telescopic view of the full Moon acquired at Lick Observatory in California. Figure (b) is outlined.

Figure 3.1(b). High-resolution Earth view. A maximum-resolution, telescopic photograph utilizing low sun illumination to maximize topographic detail. Photograph acquired at the Mt. Lemon Observatory near Tucson, Arizona. Figures (c), identified by A for Apollo, and (d), identified by O for Orbiter, are outlined.

photos was about 25 mm. The photographic images were recorded on No. 1 Isochrome film and automatically processed in a monobath solution, then washed and dried. Upon command from the ground, the negative images were scanned by a light beam created by optically focusing on the film a luminous spot from the face of a cathode ray tube. The density at each point was measured with a photomultiplier, and the signal transmitted to the Earth by means of an omnidirectional antenna. Scanning was performed across the film by horizontal movement of the spot on the cathode ray tube and along the film by a slow mechanical movement of the film.

The pictures were transmitted to the Earth shortly after the film had been processed but at a very poor signal-to-noise ratio. They were to be transmitted a second time when the vehicle was much nearer the Earth, but communication with the space station was lost and the second readout was never accomplished. The reception, which was very bad, resulted in pictures containing a great deal of noise and of poor quality. The maximum resolution of the large pictures of the Moon has been estimated to be about 16 km and the resolution of the small pictures to be about 40 km. The transmission time per picture was approximately 20 minutes; this was deduced from knowledge of the vehicle's spin rate and from counting the number of repetitive noise patterns per picture.

A map based on these pictures was prepared

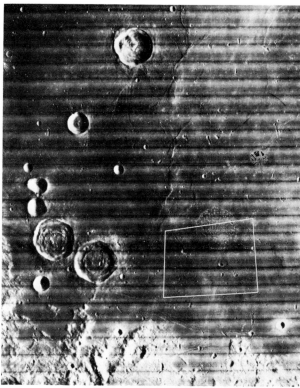

100 km

Figure 3.1 (c). Oblique view of Mare Tranquillitatis edge, with Apollo II site indicated, acquired from orbiting command ship during landing mission. Thruster in foreground.

Figure 3.1 (d). A view of the edge of Mare Tranquillitatis acquired by the high-resolution camera of Lunar Orbiter 4 (frame 85). Figure (e) outlined.

and published by the Soviet Academy of Sciences as the *Atlas of the Moon's Far Side.* Thus lunar photography from spacecraft achieved its first success and opened up new regions for exploration. A sample of the quality of these first pictures can be seen in Fig. 3.2, showing the crater Tsiolkovsky as recorded on Frame 26.

The Luna 3 mission was such a surprise to the Western world that the following spring (1960) science writer Lloyd Mallan published an article suggesting that the released lunar pictures were fakes and that the Soviets had perpetrated a Moon hoax. This was indeed a serious allegation.

The previous year Mr. Mallan had published a series of articles concerning the poor state

of Soviet science and technology and the high level of development of the Kremlin propaganda machine. These papers led to a Congressional investigation into the validity of his claims.

It must have been clear at the time to the Soviets that a hoax of the type suggested would be short-lived because the United States would soon take pictures of the back of the Moon, and any prestige temporarily achieved would be hopelessly shattered by such a disclosure. It just did not make sense. When the Luna 3 pictures are reviewed in the light of the excellent Lunar Orbiter photographs, it is seen that indeed they are honest but of poor quality. Figure 3.3 of the crater Tsiolkovsky as recorded by the low-resolution

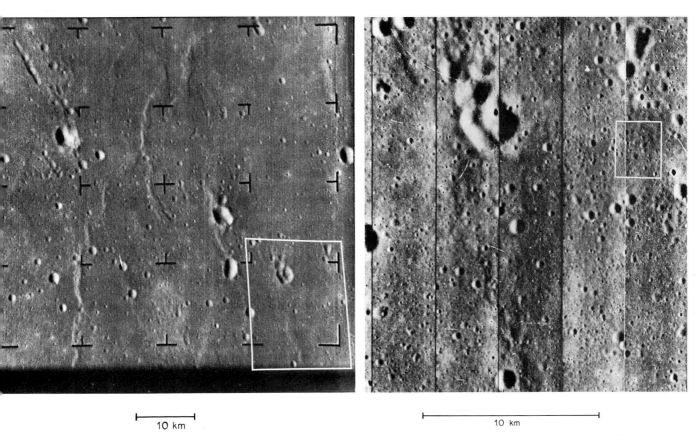

<table>
<tr><td>⊢————————⊣<br>10 km</td><td>⊢————————⊣<br>10 km</td></tr>
</table>

Figure 3.1 (e). Frame B61 of Ranger 8. Figure (f) outlined.

Figure 3.1 (f). Medium-resolution frame 75 from Lunar Orbiter 5, which flew much closer to the lunar surface than Lunar Orbiter 4. Figure (g) is outlined. Notice the abrupt change in appearance between (f) and (g).

lens (80 mm) of Lunar Orbiter 3 can be compared with Fig. 3.2 of the same area taken by Luna 3. The difference in resolution between 16 km and 300 meters is spectacular.

The existence of a large number of craters and the absence of maria on the back of the Moon were suspected from the Luna 3 pictures and confirmed first by the Soviet Probe Zond 3 and then by the U.S. Lunar Orbiter. This was one of the most important scientific surprises found on the back of the Moon. It was noted correctly by American scientists that the bright lines on the Soviet photos had been misinterpreted in the Soviet *Atlas* as the "Soviet Mountains" instead of as rays radiating from the crater.

The next Soviet photographic mission was

the Zond 3, launched July 18, 1965. This spacecraft was designed for planetary flights; similar vehicles had been sent to Mars and Venus. However, because all of these missions had been failures, it was decided to test the sensor equipment by means of a lunar flyby and retransmit the collected data from various distances to test the long-range communication systems of the spacecraft to distances as great as Mars orbit. On July 20, 1965, Zond 3 took pictures of the Moon as it passed by and successfully transmitted them to Earth from distances of 2.2 million km, 12.5 million km, and 31.5 million km. The last communication with the spacecraft was March 2, 1966, when it was 153.5 million km from the Earth, but still far short of Martian distances. Although

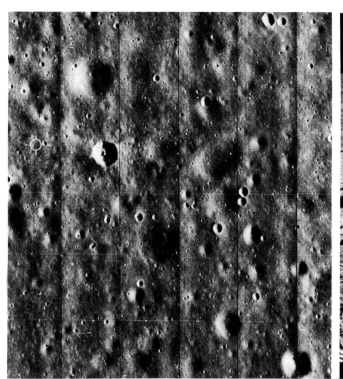

├─────────────┤
1 km

Figure 3.1 (g). High-resolution frame 76 from Lunar Orbiter 5. The Appollo 11 landing site is indicated.

Figure 3.1 (h). Astronaut Aldrin on the lunar surface at Tranquillity base, July 1969.

the lunar photography part of the experiment was successful, the spacecraft still did not exhibit sufficient reliability for a voyage to Mars.

The pictures of the Moon were taken during a 68-minute period, starting when the probe was 11,570 km from the Moon; the distance decreased to 9220 km and then increased to 9960 km, at which point the photographic period ended. The camera system used represented very little improvement over that of Luna 3, six years before.

The Zond 3 mission was designed to photograph those portions of the back of the Moon which had not been covered in the Luna 3 pictures. "Full" illumination was again used, and parts of the front of the Moon were

included in each frame so that they could be located relative to the lunar coordinate system. Since the combination of photographic distance, camera focal length, and frame size did not permit exposing the entire illuminated lunar surface on a single frame, the spacecraft was designed to perform a scanning maneuver to assure the taking of pictures of the entire area. These pictures were valuable in studies of the Orientale basin—the youngest big basin on the Moon—as this area had been seen only greatly foreshorted prior to this mission. A new atlas of the far side of the Moon was produced as a result of this new material and a map of the back of the Moon was drawn.

The Soviets continued their photographic

Figure 3.1 (i). Close-up of the lunar soil around the astronaut's foot.

Figure 3.1 (j). High-resolution view of lunar soil showing individual rock fragments.

studies of the Moon by including cameras in some of their lunar satellite missions. Few pictures and few data about the photo payload have been released. It would appear, however, that the latter is identical to that aboard Zond 3. Jodrell Bank Radio Astronomy Station in England reported receiving picture transmissions from Luna 11, although the Soviets did not mention a photographic mission for this satellite. The first Soviet press release about taking pictures from a lunar satellite described the operation of Luna 12.

Luna 12 was launched on October 22, 1966, and put into a nearly equatorial lunar orbit on October 25 with perilune of 100 km, apolune of 1740 km, and a period of 3 hr 25 min. The pictures were taken from altitudes of 100

km to 340 km. Objects as small as 15 to 20 m in size can be seen.

One release of two pictures taken by Luna 12 from an altitude of 100 km showed a 50-sq km area in the Sea of Rains. Assuming the use of a 24-mm square frame (same as Zond 3), the focal length of the lens must have been about 360 mm. The scan used was 1100 elements per line and 1100 lines per frame, the same as on the Zond 3. This would yield a resolution of about 15 m from the 100 km altitude.

There has been no official reason given for the choice of orbit or areas photographed by Luna 12. Presumably they support potential lunar landing sites of interest to the Soviets.

Early in the planning of U.S. lunar programs

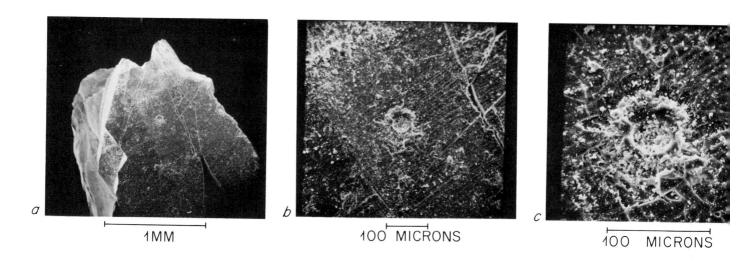

a     1MM     b     100 MICRONS     c     100 MICRONS

Figure 3.1 (k). Frames **a** through **f** show progressively greater magnification of surface detail on an individual rock fragment brought back by the Apollo 11 astronauts. The small crater seen on the surface of the fragment shown in **a** is enlarged through **d,** and

it was recognized that a great deal of information about the Moon could be acquired by a spacecraft in lunar orbit. Originally, the project suffered on two scores: the Orbiter, being of second priority, was scheduled to follow the soft lander Surveyor. Second, that lander was to be launched by the Centaur booster, which was experiencing development problems and running behind schedule. Late in 1962, the Lunar Orbiter program was given high priority and its mission was changed from general scientific exploration to specific landing-site selection in support of the Apollo mission and it was decided to use the reliable, if lower performance, Atlas/Agena launch vehicle.

Lunar Orbiter spacecraft were ejected from parking orbit to a trajectory past the Moon. They used three-axis stabilization similar to that of the Mariners and had propellants for two midcourse corrections. As they approached the Moon, a retroengine first put them into a circular orbit and then into an elliptical one, depending upon the particular objectives of the mission.

The photographic payload weighed about 60 kg and was enclosed in a shell maintained at 1.7 psi internal pressure. Two windows were provided for the camera lenses. The two cameras, one with an 80-mm f/5.6 lens and the other with a 610-mm f/5.6 lens, took pictures alternately on a single roll of 70-mm film. (This is reminiscent of the Luna 3 cameras, which took two pictures on a single strip of 35-mm film.) Image motion compensation was included in the camera and controlled by an instrument which measured the velocity-to-height ratio of the spacecraft. The format for the 610-mm lens was 219 mm by 55 mm; for the 80-mm lens, 65 mm by 55 mm. Exposures of 0.04, 0.02, or 0.01 sec were used, depending upon the lighting. Eastman SO-243-type film was used, a fine-grain, high-contrast film with an Aerial Exposure Index of about 3. The exposed film was processed automatically inside the pressure shell using Bimat Transfer Film.

The readout system incorporated a line-scan tube which was a further development of one that had been used successfully in previous

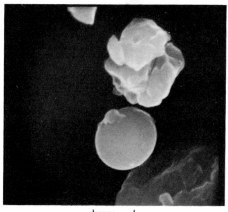

10 MICRONS    *d*      1 MICRON    *e*      1 MICRON    *f*

then individual particles are enlarged further. This series of pictures was acquired with the scanning electron microscope by John Devaney of the Jet Propulsion Laboratory and Gerald Wasserburg of Caltech.

space operations. This tube contains a phosphor-coated drum which spins during readout. A very bright moving spot is created by scanning the spinning drum axially with an electron gun. The optical-mechanical scanner used this moving-light source together with mechanical movement of the film. The scanner lens reduced the spot size to 6.5 microns at the film plane. A photomultiplier was used to measure the light output and generate the video signal, which was amplified and then transmitted to Earth.

Five Lunar Orbiter spacecraft were launched between July 1966 and August 1967, and all were successful. Altogether, about 815 useful frame pairs of pictures were returned. This represented a tremendous amount of data, as the resolution was 75 to 100 line pairs per millimeter. The first three spacecraft were placed in elliptical orbits with perilune 50 km above the lunar surface to obtain high-resolution pictures of the potential Apollo landing sites. From this altitude, photographs were obtained with resolutions better than one meter per line pair. Since there were no

spacecraft failures, the last two orbiters were programmed to complete the reconnaissance photography of the entire lunar surface at excellent resolution. During its mission the fourth Lunar Orbiter photographed about 60 percent of the Moon's surface at a resolution of approximately 60 m, which is about ten times better than that obtainable with Earth-based telescopes (see Appendix C, Fig. C.4).

Thus film readout has provided most of the scientific photography for the Moon. It has been demonstrated to be a superior—a classy—technical approach for space photography. On the other hand, if the space vehicle is going to be returned safely to Earth anyhow, then why not wait and develop the film in a conventional photo lab here?

### 3.3 FILM RETURN—BRING IT BACK ALIVE
From the point of view of a photographer, an exciting facet of the manned lunar programs is that they require the physical return of the astronaut and incidentally of the film. If the primary role of the mission were solely to bring back pictures, it could be accomplished

43

Table 3.1 Characteristics of photographic cameras for Moon missions

| Mission camera | Lens focal length (mm) | Focal ratio (f/no.) | Format (cm) | Number of pictures/magazine |
|---|---|---|---|---|
| Zond 6 | 400 | 6.3 | 13 × 18 | 200 |
| Apollo Hasselblad | 80 | 2.8 | 6 × 6 | 200 B&W, thin base |
| | 250 | 5.6 | | 160 color, thin base |
| | | | | 100 standard base |
| Apollo Itek Panoramic | 610 | 3.5 | 11 × 113 | 1490 |
| Apollo Hycon Frame | 450 | 4 | 11 × 11 | 480 |
| Apollo Fairchild Mapping | 75 | 4.5 | 11 × 11 | 3600 |

### 3.2 Luna 3 picture of the crater Tsiolkovsky

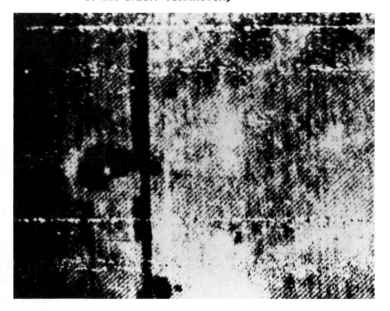

This large crater with a central peak, located on the far side of the Moon, was first photographed by the Luna 3 spacecraft. Although the quality of the picture is poor, the characteristic shape of the crater is easily discerned.

with a smaller, less expensive spacecraft. Unmanned photographic return missions have been adopted, but only incidentally, in the Soviet lunar program and not at all by the United States. Unmanned roundtrip missions for Mars will become feasible for scientific exploration long before any manned adventures in the direction of that planet are possible.

The Soviet Zond 5, 6, and 7 missions appear to have been tests for manned flights around the Moon, similar to the Apollo 8 manned flight of December 1968, except that they were probably not intended to go into and out of orbit about the Moon. Thus their photographic capabilities must be thought of as secondary objectives, as are those for the Apollo flights. One of the curious aspects of the Apollo mission has been that the primary camera has been a hand-held, modified, commercially designed Hasselblad camera commonly used by many amateur and some professional photographers. Reconnaissance pilots in aircraft gave up the use of hand-held cameras in World War I, and since that time specially designed aerial cameras have been in use. Today these cameras are designed for specific missions and emphasize particular photographic characteristics such as high resolution, wide angle, minimum geometric distortion, and others. Thus it might be inferred that the use of a hand-held Hasselblad in the Apollo program reflects the low priority of lunar pictures for scientific study. On the other hand, later Apollo missions are scheduled to carry excellent high-quality aerial cameras to obtain high-resolution lunar photographs and to take mapping pictures with precise geometric use. (The ill-fated Apollo 13 carried advanced photographic equipment.) Pictures from these cameras will be very useful in viewing detail over large lunar distances, in measuring distances between landmarks on the lunar

### 3.3 Lunar Orbiter picture of the crater Tsiolkovsky

This low resolution picture, taken by Lunar Orbiter 3, is typical of the pictures taken by these spacecraft of the entire lunar surface. The quality is much better than is possible to obtain using a telescope on the Earth's surface. The accompanying high resolution pictures recorded fine topographic detail over large areas and will constitute the source material for lunar studies for a long time.

surface, and in determining the shape of the Moon.

The Soviet probes Zond 5, 6, 7, and 8 went around the Moon and returned to Earth; each carried a small aerial mapping camera, remotely operated, and programmed to take three successive frames at different settings of the diaphragm to assure proper exposure. The camera was carefully calibrated photometrically and photogrammetrically before flight. Presumably Zond 6 and 7 carried the same camera; the most conspicuous difference between the published pictures was that Zond 7 obviously carried color film. Table 3.1 summarizes some of the characteristics of these cameras. Figure 3.4 is a sample of the lunar photography taken by Zond 7.

Because of the great difference in style between missions which physically return film and those which transmit pictures by telemetry, it is important to examine the characteristics and advantages of film return, as these missions will probably always be more expensive than readout. Film is an excellent storage mechanism for picture bits (see Appendix A). Because most cameras can cycle very rapidly, it is possible to take a large number of frames in a short time and thus rapidly store a huge number of bits. The weight of film is small compared to its storage capability.

When compared to vidicon tubes, the film format is very large and the packing density great. When comparing space-rated systems, film should run from 50 to 150 line pairs

### 3.4 The Moon—a Soviet view

This image of the northwest boundary of Oceanus Procellarum is a copy of the color picture acquired by the Soviet flyby and return unmanned spacecraft Zond 7 in August 1969. The picture was taken remotely at a distance of several thousand kilometers and the exposed negative returned directly to the Soviet Union in an instrument capsule.

45

per millimeter, whereas vidicon tubes might achieve 30 to 50 line pairs per millimeter. Although the format and resolution of film readout pictures might be comparable to that of returned film, automatic processing in the spacecraft must be substituted for laboratory processing and there is always degradation of the image in the scan mechanization.

Certainly one of the greatest advantages of film return is in photogrammetry for the producing of topographic maps. Here the large format at high resolution is required to draw accurate contours over large areas; of course the pictures should also be essentially free from geometric distortion, which is difficult to achieve from readout materials. Film is the best material whenever there is need to make geometric measurements from pictures; measurements are used for the computation of a geodetic control net, determination of planetary radii, solving navigation problems, and more. When the lunar pictures from the Apollo advanced cameras are returned, Man's knowledge of the details of the entire surface of the Moon will be greatly increased.

## 3.4 MAPPING AND NAMING THE FEATURES OF THE MOON

The subject of selenography—the study, description, and cartographic representation of the Moon's surface—began in 1610 when Galileo described the appearance of the lunar surface as viewed through a telescope. With the introduction of the telescope, our knowledge of the Moon underwent a revolutionary development. The Moon's rocky surface features were revealed, and the absence of many Earth-like characteristics such as oceans, seas, clouds, and rivers surprised the early astronomers. The telescope was the major instrument for lunar exploration for over three hundred years until the introduction of the unmanned spacecraft in the late fifties. Recently, samples of the Moon have been brought to Earth, heralding yet another quantum jump in lunar exploration. Thus there have been four historical periods in lunar exploration: the eyeball age, the telescope age, the robot age, and the sample-return age. In terms of best resolution available at the lunar surface, the human eye sees the Moon at a resolution of about 120 km, the telescope is limited by the Earth's atmosphere to a resolution of about a kilometer, and unmanned space photography increased that to a meter. In the first two historical periods, little over one-half of the surface of the Moon was available for observation from the Earth since the same side of the Moon always faces the Earth (except for librations). For the first time in history the entire globe is now open to man's scrutiny.

But even long before the advent of selenography, the Greek philosophers had speculated about the nature of the Moon and its surface. Many of their ideas survived and influenced the writings of medieval scholars. However,

46

most of the astronomy texts published before 1600 dealt with the motions of the planets and carefully avoided describing their nature. In fact, it is surprising that no drawings of the lunar surface from that period have been published or have survived. In the astronomy books of that time, the Moon commonly bears a man's face, as it still does in our popular almanacs. Although Leonardo da Vinci (1452-1519) is not usually considered an astronomer, he was an excellent observer and scientist, and many of his sketches and notebooks are available today. His drawings of the Moon, made without the aid of a telescope, are mentioned in his notes but have apparently been lost.

During the fifteenth and sixteenth centuries a variety of theories attempted to account for the appearance of the lunar surface. Perhaps the most popular explanation was that the lunar surface was smooth and polished like a mirror and thus reflected an image of the Earth. Da Vinci recognized the fallacy of this explanation and wrote:

the Earth, when not covered by the water, presents different shapes from different points of view; so when the Moon is in the east, it would reflect other spots that when it is overhead or in the west, whereas the spots upon the Moon, as seen at full Moon, never change.

Thus, he noticed that the Moon always presented the same face to the Earth. He continues:

A second reason is that an object reflected in a convex surface fills only a small part of the mirror, as is proved in perspective. The third reason is that when the Moon is full it only faces half the orb of the illuminated Earth.

The notion that the lunar surface was mirror-like must have been very popular; Galileo refers to it in 1610 when first describing the surface of the Moon as viewed with a telescope:

the surface of the Moon is not perfectly smooth, free from inequalities and exactly spherical, as a large school of philosophers considers with regard to the Moon and the other heavenly bodies, but that, on the contrary, it is full of inequalities, uneven, full of hollows and protuberances, just like the surface of the Earth itself, which is varied everywhere by lofty mountains and deep valleys.

Years later (1632) he again presented arguments why the surface of the Moon could not be smooth like a convex mirror, indicating that such notions were still popular even after the invention of the telescope.

Galileo was the first to use the telescope for astronomical research and to publish his results. The impact was rapid and the age of the telescope had begun. When Galileo first viewed the Moon, he was surprised at the roughness of the surface and mountains. The plains (dark areas) he mistook for seas:

if anyone wished to revive the old Pythagorean opinion that the Moon is like another Earth, its

47

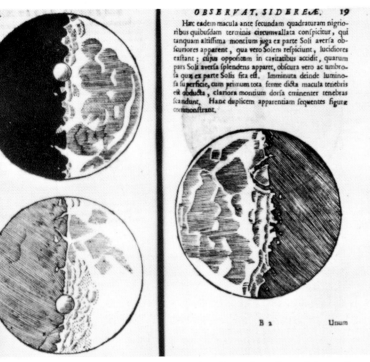

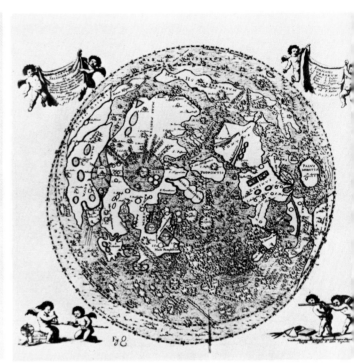

The first drawings of the Moon based on telescopic observations were made by Galileo and published in 1610. Although craters are shown, the quality of the woodcuts is very poor.

In 1647 Hevelius published **Selenographic,** a beautiful and comprehensive study of the Moon which contained many illustrations. Only ten of the 250 names of lunar features proposed by Hevelius are in common usage today.

brighter part might very fitly represent the surface of the land and its darker region that of the water. I have never doubted that if our globe were seen from afar when flooded with sunlight, the land regions would appear brighter and the watery regions darker.

Why did Galileo misinterpret the Moon's dark areas? How good was his telescope? Recently two of his telescopes were removed from the museum in Florence and tested on an optical bench. The resolving power of the instruments turned out to be about 10 seconds of arc, which at lunar distance corresponds to a surface resolution of about 15 km—not much different from the Luna 3 pictures of

the back of the Moon. It is not surprising that there were some errors in interpretation.

Galileo published in 1610 five drawings of the Moon, and all of these woodcuts are of very poor quality; three of them are shown in Fig. 3.5. It is regrettable that these were not drawn and reproduced with more care; features cannot be identified except, perhaps, for the large crater (Ptolemaeus).

A number of maps were drawn as the use of the telescope become popular. Most of these were better than the drawings of Galileo. In 1647 Johannes Hevelius (1611–1687), the city counselor of Danzig, published *Selenographic,* which contained many sketches of the

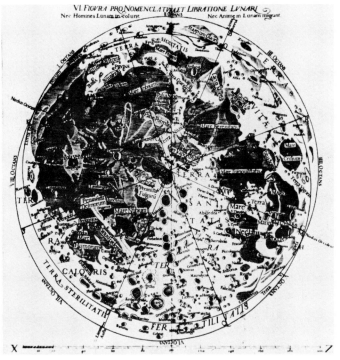

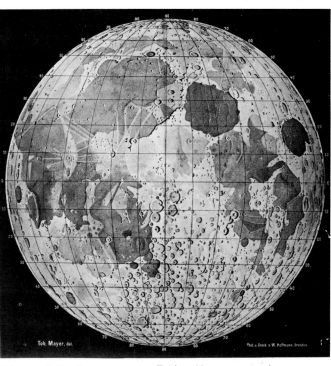

In 1651 Riccioli published this map which was prepared from observations made by his friend Grimaldi. Most of the names for lunar features proposed by Riccioli have been accepted by the IAU.

This carefully drawn map by Tobias Mayer was published thirteen years after his death in 1762. The original has a nineteen centimeter diameter and was the only reliable lunar chart for fifty years.

Moon, a map, and a description of his lunar work, which included measuring the heights of mountains. Figure 3.6 from this book shows the beautiful quality of the copperplates. This became the standard reference work for many years. Hevelius gave names to about 250 lunar features, only ten of which, however, are in use today. The nomenclature suggested by Giovanni Baltista Riccioli (1598–1671), still in general use today, was contained on a map, published in 1651 and drawn by his friend Francesco Grimaldi (1618–1663), also of Bologna University. This map can be seen in Fig. 3.7.

The next major step in lunar mapping was

taken by Tobias Mayer (1732–1762) of Gottingen, who used cartographic techniques to locate features with precision. Using a micrometer, he measured the position of 24 points and located an additional 65 points with respect to the initial ones. He determined the Moon's axis of rotation and the inclination of the equator. None of his maps were published during his lifetime, but his 19-cm diameter map, published in 1775 (see Fig. 3.8), was considered for fifty years to be the most accurate representation. It is interesting to note that after this period it became customary to place south at the top of the page, as it appears in the telescope. The earlier maps

49

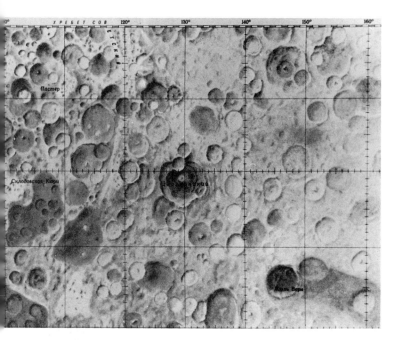

3.9a. Reproduced in black and white is a portion of the 1:5,000,000 lunar chart prepared by the Sternberg Astronomical Institute, Moscow University, in 1966. It is clear that topographic details in this region were vague at that time.

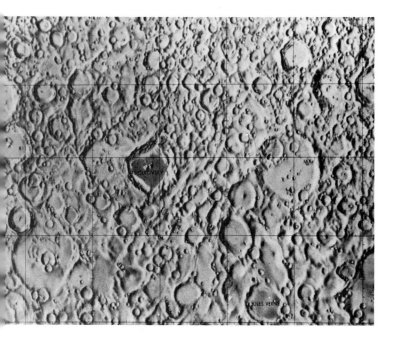

3.9b. Reproduced in black and white is a portion of the 1:5,000,000 lunar chart prepared by the USAF Aeronautical Chart and Information Center, St. Louis, in 1967. The Lunar Orbiter photography permitted detailing many little-understood areas of the Moon.

(that is, Hevelius and Riccioli) had north at the top, no doubt owing to the types of eyepieces used in the telescopes of the day. South continued at the top until the International Astronomical Union in 1961 recommended that maps for exploration purposes be printed with north on top.

The next technological step was the use of photography in the last half of the nineteenth century to take beautiful pictures of the Moon with the various existing telescopes. These efforts led to the publication of a number of photographic atlases and permitted great precision in locating features by measuring the location of points on photographic plates. J. H. Franz in Germany and S. A. Saunders in England used this technique to locate about 3000 points with an average accuracy of 300 meters. Finally, in the late 1950s, G. P. Kuiper and his associates produced a comprehensive series of lunar atlases.

An ambitious modern mapping project was undertaken in the early sixties by the Aeronautical Chart and Information Center (ACIC) of the U.S. Air Force to produce charts of the Moon at a scale of 1:1,000,000. It would require 144 charts of this LAC series to cover the Moon. Forty-four charts covering most of the front of the Moon have been produced, but the project has been terminated for the present.

ACIC has produced maps of the entire lunar surface at scales of 1:5,000,000 (3 charts) and 1:10,000,000 (1 chart) based on the

Lunar Orbiter pictures of the back of the Moon. The Soviet Union produced a map at a scale of 1:5,000,000 (7 charts) in which the features of the back of the Moon were taken from the Luna 3 and Zond 3 photography. Figure 3.9 shows the crater Tsiolkovsky as it is portrayed on these different charts.

As mentioned previously, lunar nomenclature was the subject of confusion for the first century or so following the introduction of the telescope. With the naming of features, the science of selenography continues to reflect the times and the emotions of the personalities involved. In 1935 the International Astronomical Union adopted the names of the lunar features based on the system of Blagg and Muller. The lunar maria are named for moods or states of mind, lunar mountain ranges after terrestrial mountain ranges, and lunar craters after people. Less prominent features use the name of a nearby feature followed by one or two letters. Shortly after the start of the space age, new areas of the Moon were viewed for the first time and the naming game started over. In 1961 the IAU adopted a list of names proposed by the Soviet delegation for features discovered by Luna 3. Another group of Soviet names, based on the Zond 3 pictures was proposed at the IAU meeting in 1967, but was not officially adopted. However, at the 1970 meeting a list of more than five hundred new names recommended by the nomenclature committee was adopted.

Thus the names of lunar features, and eventually those of the planets as well, constitute a permanent historical record of the changing cultural backgrounds of the scientists and their nations involved in lunar and planetary exploration.

**REFERENCES**

Antonov, S. M., K. S. Bogomolov, N. I. Kirillov, N. S. Ovechkis, and V. I. Uspenskii, "Photographic Processes Used in the First Photographs of the Other Side of the Moon," *Artificial Earth Satellites,* Vol. 9, July 1962.

Elle, B. C., C. S., Heinmiller, P. J. Fromme, and A. E. Neumer, "The Lunar Orbiter Photographic System," *Journal of the SMPTE,* Vol. 76, No. 8, August 1967.

Heacock, Raymond L., "Lunar Photography: Techniques and Results," *Space Science Reviews,* Vol. 8, No. 2, April 1968.

Katz, Amrom H., "Analysis of Lunick III Photographs," *Proceedings of the Lunar Planetary Exploration Colloquium,* Vol. II, No. 2, March 17, 1960.

Lipskii, Yu. N., "Points From a Study of the First Photographs of the Reverse Side of the Moon," *Planetary and Space Science,* Vol. 9, September 1962.

McCauley, John F., *Moon Probes,* Little, Brown and Company, 1969.

Whitaker, E. A., "Evaluation of the Russian Photographs of the Moon's Far Side," *Communications of the Lunar and Planetary Laboratory,* The University of Arizona, Vol. 1, No. 13, 1962.

"*Three ships is a lot of ships. Why can't you prove the world is round with __one__ ship?*"

# CHAPTER 4

# Distance and ignorance
# in the exploration
# of the planets

## 4.1 THE INVERSE SQUARE DEPENDENCE OF KNOWLEDGE ON DISTANCE

The most important obstacle to exploration of the planets is their distances from the Earth. Ground-based observations, which necessarily supply the a priori knowledge upon which space probes are conceived and flown, are degraded by the progressively lower surface resolution for each more distant object. Thus the smallest area resolvable from Earth increases in proportion to the square of the distance from Earth, as illustrated in Table 4.1 and Figure 4.1. The space probes themselves require not only a progressively greater launch vehicle and lifetime capability, but also the communication requirements increase in proportion to the square of the distance to the Earth. Thus, both a priori knowledge and space probe communication capability can be considered to diminish inversely with the square of the distance to the planets from the Earth.

In reality, the effects of distance are even more serious in that the planets can be observed from the ground only at very restricted times and under particular conditions of illumination. The background of ground-based observations concerning the planets necessarily contains extreme seasonal, orbital, longitudinal, and latitudinal selection effects. Other kinds of physical observations, such as radio emissions, are equally degraded by the inverse square effect of distance. Radar reflection measurements suffer by the inverse fourth power of range, and also are influenced by the large variation in rotational speeds of the terrestrial planets.

A further difficulty in exploration of the planets compared to the Moon is that most have surface features which are temporally variable. For example, Mars exhibits daily, seasonal, and irregular variations in surface markings and atmospheric features. It's surface is relatively highly colored as well. Time and color variations are characteristic of the giant planets and they exhibit enormous surface areas as well. Both color and temporal variations of surface and atmospheric features increase substantially the relative amount of photography necessary to obtain comparable understanding of the surface or atmosphere. It thus becomes obvious that (1) our a priori information about the planets is enormously less than was the case for the Moon before space flights began; (2) except for Mercury and the Galilean satellites, planetary objects will require much larger amounts of photography, over extended periods of time, as well as many other close-up observations for even preliminary exploration than has been the case for the Moon; and (3) the rate of acquiring new, close-up information will be low due to the great cost and difficulty of returning data from great distances to the Earth.

**Table 4.1 Relative knowledge of the planets from telescopes (compared to Moon)**

| | Minimum distance from Earth, kms | | Maximum useful surface resolution km** | | Area = $(\pi)$ (mean D)$^2$ km$^2$*** | |
|---|---|---|---|---|---|---|
| Moon | $3.7 \times 10^5$ | (1) | $9.3 \times 10^{-1}$ | (1) | $1.9 \times 10^7$ | (1) |
| Mercury* | $1.2 \times 10^8$ | $(3.3 \times 10^2)$ | $3.1 \times 10^2$ | $(3.3 \times 10^2)$ | $8.3 \times 10^7$ | (3.6) |
| | $8.2 \times 10^7$ | $(2.2 \times 10^2)$ | | | | |
| Venus* | $8.3 \times 10^7$ | $(2.2 \times 10^2)$ | $2.1 \times 10^2$ | $(2.2 \times 10^2)$ | $5.0 \times 10^8$ | $(2.2 \times 10^1)$ |
| | $4.3 \times 10^7$ | $(1.2 \times 10^2)$ | | | | |
| Mars | $5.5 \times 10^7$ | $(1.5 \times 10^2)$ | $1.4 \times 10^2$ | $(1.5 \times 10^2)$ | $1.5 \times 10^8$ | $(6.5 \times 10^0)$ |
| Jupiter | $6.7 \times 10^8$ | $(1.8 \times 10^3)$ | $1.7 \times 10^3$ | $(1.8 \times 10^3)$ | $6.1 \times 10^{10}$ | $(2.6 \times 10^3)$ |
| Saturn | $1.2 \times 10^9$ | $(3.4 \times 10^3)$ | $3.1 \times 10^3$ | $(3.4 \times 10^3)$ | $4.2 \times 10^{10}$ | $(1.8 \times 10^3)$ |
| Uranus | $2.6 \times 10^9$ | $(7.0 \times 10^3)$ | $6.5 \times 10^3$ | $(7.0 \times 10^3)$ | $7.0 \times 10^9$ | $(3.1 \times 10^2)$ |
| Neptune | $4.4 \times 10^9$ | $(1.2 \times 10^4)$ | $1.1 \times 10^4$ | $(1.2 \times 10^4)$ | $6.3 \times 10^9$ | $(3.3 \times 10^2)$ |
| Pluto | $4.6 \times 10^9$ | $(1.3 \times 10^4)$ | Unresolved | | Indeterminate | |

*The upper figure refers to the distance of the planet at 40 percent illumination. This result is used in column two. The lower figure is the minimum distance from the planet to Earth, or at an inferior conjunction.

## 4.2 THE SIGNIFICANCE OF A PRIORI IGNORANCE

We have just emphasized the importance of a priori knowledge of the surface to be explored; in Appendix A this concept is pursued further. In this section we extend this line of reasoning to point out that definite, but incorrect, viewpoints held by the experiment designers and interpreters can complicate or inhibit the process of discovery when an unknown surface is being explored. Indeed, an essential element of exploration, especially in the early phases, is to reveal the magnitude of our previous ignorance by eliminating hypothetical configurations of the unknown surface which had come to be regarded as possible or even probable.

Man's first close-up look at Mars in 1965 with the U.S. probe Mariner 4 was a striking

**4.1 A priori ignorance**

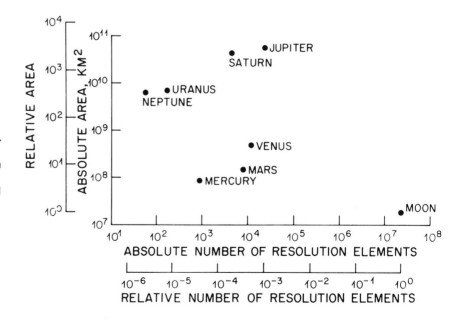

Figure 4.1 (a) **One quantitative view.** In this diagram the total surface area (vertical axis) is compared with a measure of total telescopic information about that planet (horizontal axis). Table 4.1 supplies the detailed values. The axes are also displayed in normalized units compared to the Moon. Inasmuch as this is a log/log plot, the enormous differences in surface area and a priori knowledge are especially evident.

| | Number resolution elements | | Number resolution elements per km² | | resolution element km² per | |
|---|---|---|---|---|---|---|
| Moon | $2.2 \times 10^7$ | (1) | 1.2 | (1) | .87 | (1) |
| Mercury* | $8.8 \times 10^2$ | $(3.3 \times 10^{-5})$ | $1.0 \times 10^{-5}$ | $9.1 \times 10^{-6}$ | $9.6 \times 10^4$ | $1.1 \times 10^5$ |
| Venus* | $1.2 \times 10^4$ | $(4.4 \times 10^{-4})$ | $2.3 \times 10^{-5}$ | $2.0 \times 10^{-5}$ | $4.4 \times 10^4$ | $5.1 \times 10^4$ |
| Mars | $8.0 \times 10^3$ | $(2.9 \times 10^{-4})$ | $5.0 \times 10^{-5}$ | $4.6 \times 10^{-5}$ | $2.0 \times 10^4$ | $2.2 \times 10^4$ |
| Jupiter | $2.4 \times 10^4$ | $(8.8 \times 10^{-4})$ | $3.5 \times 10^{-7}$ | $3.1 \times 10^{-7}$ | $2.9 \times 10^6$ | $3.3 \times 10^6$ |
| Saturn | $4.3 \times 10^3$ | $(1.6 \times 10^{-4})$ | $1.0 \times 10^{-7}$ | $8.3 \times 10^{-8}$ | $9.6 \times 10^6$ | $1.1 \times 10^7$ |
| Uranus | $1.7 \times 10^2$ | $(6.2 \times 10^{-6})$ | $2.4 \times 10^{-8}$ | $2.0 \times 10^{-8}$ | $4.2 \times 10^7$ | $4.8 \times 10^7$ |
| Neptune | $5.8 \times 10^1$ | $(2.1 \times 10^{-6})$ | $8.3 \times 10^{-9}$ | $6.9 \times 10^{-9}$ | $1.2 \times 10^8$ | $1.4 \times 10^8$ |
| Pluto | Indeterminate | | Indeterminate | | Indeterminate | |

**0.5 arc second is considered to be nominal maximum resolution for ground-based telescopes.

***The area for the Moon used in the table is ½ the actual area.

demonstration of how far from reality the scientific "climate of opinion" can be. The incorrect prejudices that had developed about both the surface and atmosphere of Mars over the previous 75 years were so firmly held by almost all scientists concerned with the planet that, had the initial U.S. and Soviet lander attempts been successfully prosecuted, they would have certainly resulted in ignominious failures, especially because of the erroneous belief that the atmosphere pressure was only about 1/10 that of the Earth. Similarly, the designers of the Mariner 4 television experiment were unprepared for the existence of a lunar-like landscape on Mars. That mission was of such limited capability, however, that there was little opportunity for their a priori ignorance to affect the experiment design.

There are some similarities in the exploration of Venus. The designers of the Soviet entry

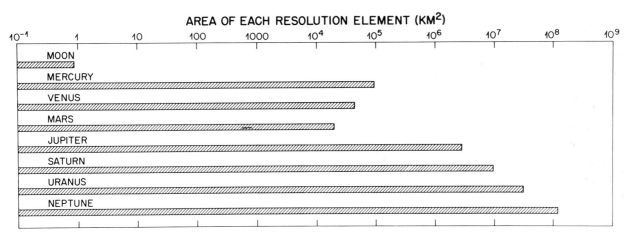

**Figure 4.1 (b). Another measure.** This bar graph shows on a log scale the surface area on each planet under best conditions represented by the limiting resolution from Earth. Hence the smallest resolvable area on Mercury is 100,000 times larger than that of the Moon yet still 20 times smaller than that of Jupiter or the Galilean satellites.

55

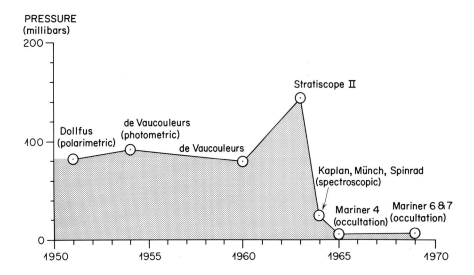

PRESSURE
(millibars)

Various estimates of the atmospheric pressure on Mars are plotted according to time of publication. It can be seen that different observers using different techniques inferred estimates that were uniformly an order of magnitude too large throughout the 1950s, further contributing to an (unjustified) similarity to Earth.

capsule known as Venera 4 apparently thought beforehand that Venus had an atmosphere through which this craft could survive to the surface; it did not. Only after Venera 5 and 6 also failed to reach the surface was the extremely dense atmosphere of the planet acknowledged.

More significant, however, is the fact that the initial (and continuing) U.S. exploration strategy has placed primary emphasis on Mars. This priority reflected early assumptions concerning the presumed similarity of Mars and the Earth. This early misconception slowed the development of efficient exploration approaches, and delayed the acquisition of many new observations.

Because many of the misinterpreted observations were based on visual and photographic imagery acquired with ground-based telescopes, it is particularly appropriate to digress at this point to investigate just how the scientific community became so misled about Mars. Discovery ultimately takes place in the minds of men; it is important to understand reasons for failure as well as success so that our knowledge of other worlds can increase in proportion to our capability to acquire new information about them.

Mars was first studied seriously through the telescope in the middle of the nineteenth century. The most striking features were the bright polar caps which regularly developed in the winter hemisphere and the seasonal changes in the outline of the dark markings. Both seemed to be characteristic of how the Earth was imagined to look if viewed from space, assuming the changes in markings reflected seasonal vegetative changes. In addition, Mars has almost the same obliquity (tilt of the axis of rotation relative to the plane of the ecliptic) as has the Earth and the length of a Martian day is within forty minutes of that of the Earth. Thus it was natural for Mars to be regarded as a close relative of the Earth, even though Mars is closer in size to Mercury than to the Earth (see Fig. 6.2). Indeed, Venus is the twin of the Earth from the point of view of size.

It was apparent very early in the investigations that Mars has an atmosphere because cloud-like features, "dust storms," and such, were observed. Nothing was known of its atmospheric composition, however, until the late 1940s, when $CO_2$ was identified in an amount some thirty times that of the Earth. Various attempts were made to estimate atmospheric surface pressure based on indirect effects. These estimates tended to cluster around a value 1/10 that of the Earth until as recently as the middle 1960s, as shown in Fig. 4.2. Because oxygen, nitrogen, and water are all difficult to detect through the Earth's atmosphere, it could be presumed when the first Mars probes were planned that Mars had a somewhat thinner but still Earth-like atmosphere, and exhibited mountains, ancient ocean basins, and possibly even pools of liquid water in some places.

We now know that the surface pressure is more like 1 percent rather than 10 percent that of the Earth, that $CO_2$ is the principal constituent of the atmosphere, and that water vapor is exceedingly rare, less than 1/1000 that of the Earth. Furthermore, liquid water can hardly exist at all at the surface since the *total* atmospheric pressure barely exceeds the triple point of water. The frost caps are carbon dioxide frost, not water ice. Plant life like that of the Earth simply cannot survive in such a hostile environment; those interested in the possibility of life on Mars are now concerned with thinly distributed, highly specialized micro-organisms that might have adapted from some hypothetical earlier time when the environment was more favorable. Finally, the surface topographically resembles that of the Moon rather than the Earth; there is no evidence of the mountain-building processes characteristic of Earth. Thus, there now seems little reason to have singled out Mars as an especially promising site for extraterrestrial life; in any case, the supposed similarity to Earth has vanished in the face of better information.

As Mars still was believed to be similar to the Earth until the 1960s, we can envision the "climate of opinion" which prevailed in the latter part of the nineteenth and early twentieth centuries when the dominating personality of Percival Lowell imprinted itself on the subject of Mars. Lowell was one of the first astronomers to appreciate the importance of site location in obtaining good imagery with ground-based telescopes. He established an observatory at Flagstaff, Arizona, which probably provided the best views of Mars in the world at that time, and he pursued the observation program with dedication from the end of the nineteenth century until the First World War. Yet he not only drew incorrect conclusions, but ultimately produced biased and misleading observations. The controversy surrounding him and his notion that Mars was inhabited by intelligent beings dominated the subject of planetary investigations for half a century. Indeed, a long hiatus in careful evaluation of Martian observational data is one of the principal reasons some modern scientists had uncritically accepted the existence of plant forms on Mars; in fact these

57

Map I. Beer and Maedler, 1840.

The following figures record how man's view of Mars has changed, first from the crude (but rather faithful) early work, through Lowell's progressively straighter canals, to a recent telescopic map, and finally to a photomosaic developed from 1969 spacecraft pictures. The overemphasis on "canals," and the unduly sharp boundaries erroneously inferred by many telescopic observers are apparent. Figures 4.3(a) through 4.3(f) are reproduced directly from Lowell's original work. Figure 4.3(g) is a high contrast version of the planning chart used for the Mariner 6 and 7 flybys and depicts how the planet was expected to appear for them. Figure 4.3(h), also with contrast exaggerated, is a mercator mosaic of full disc pictures acquired by the two spacecraft, and involved very little subjective human interaction in its preparation.

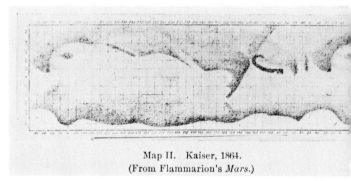

Map II. Kaiser, 1864.
(From Flammarion's *Mars*.)

(a)

traditional views had been part of "the climate of opinion" for so long that they were not questioned.

How could this happen? What drove Lowell to destroy himself as a scientist and, in the process, practically destroy the subject itself? Is this a case of an intellectual gold bug? The answer seems to be a classic case of an idea overcoming the man. Scientists are hunters of ideas. But sometimes the hunter becomes the hunted if he encounters an idea too strong for him. This is a human process, probably as prevalent today as in Lowell's time, and therefore it is a potential hazard for all those who would explore other worlds.

At the time Lowell first began his observations, Schiaparelli had identified some very faint, dark linear features observable at some time on Mars. Four of his maps are shown in Fig. 4.3. He referred to these as *canali* (meaning channels or grooves) and they came to be known as "canals." Lowell became intrigued with these canals and, as shown by the chronological series of his own maps shown in Fig. 4.3, his rendition of the canals became progressively straighter, then great circles, and finally showed intersections as "oases," suggesting, to him, intelligent civilization. Figure 4.3(h) shows how the planet actually looks photographicallly in the same format.

This, of course, is all nonsense and was a result of an overheated imagination which fogged his vision. But he could imagine no explanation for the seasonal changes other than vegetation on what seemed to be an Earth-like planet. Indeed, there still is no

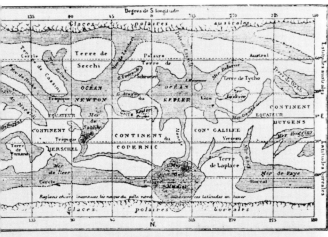

Map III. Résumé by Flammarion, 1876.
(From Flammarion's *Mars*.)

Map IV. Green, 1877.
(From Flammarion's *Mars*.)

(b)

Map V. Schiaparelli, 1877.
(From Schiaparelli's *Memoria*.)

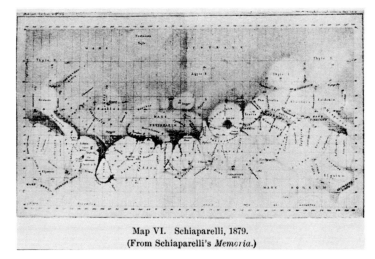

Map VI. Schiaparelli, 1879.
(From Schiaparelli's *Memoria*.)

(c)

satisfactory explanation for this phenomenon although a biological explanation seems less likely now that we know more about the appearance of the Earth from space as well as about the hostile surface of Mars. And, he might have reasoned, if plants, why not animals. And, if animals, why not men? Intelligent life on Mars could not have been rigorously disproved at the turn of the century nor were satisfactory alternative explanations of either the seasonal variations or "canals" available. Actually, intelligent life on Mars was not a bad hypothesis then when *compared to the alternatives*. This is the key point. What Lowell failed to appreciate is that any list of alternative explanations of the observations, even today's list, can never be complete. The unimagined explanations often prove to

be the correct ones. Hence, the fact that a particular explanation is the least unsatisfactory one that has been suggested is not particularly significant. Similarly, the profound importance of photographic exploration is that it provides the opportunity to discover situations which were *unimaginable beforehand*. This distinction lies at the heart of the differences between exploration and experimentation, and is an important criterion in assessing the relative importance of various instruments proposed for space missions.

To return to the development of ideas about Mars, Lowell was not willing to accept a small but finite possibility of intelligent life on Mars as the least unsatisfactory explanation of the observations. The appeal of that possibility was too great for him to let the

59

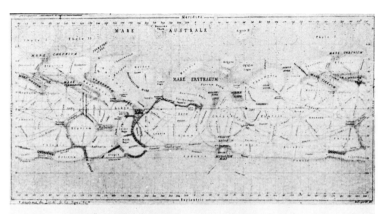

Map VII. Schiaparelli, 1881.
(From Schiaparelli's *Memoria*.)

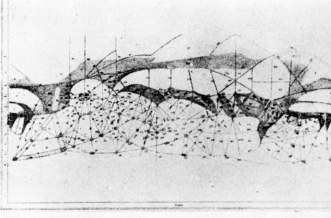

Map IX. Lowell, 1894.

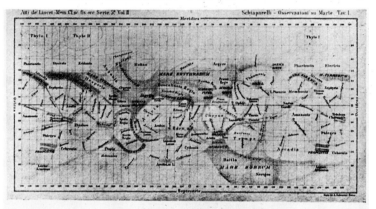

Map VIII. Schiaparelli, 1884.
(From Schiaparelli's *Memoria*.)

(d)

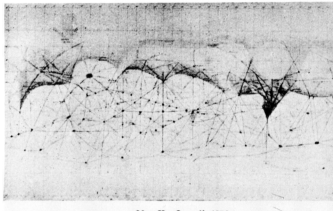

Map X. Lowell, 1896.

(e)

matter be so uncertain. Thus, he began to imagine that the *canali* were actually canals, and drew them accordingly. Like many other human beings in more personal matters, he could not stand the uncertainty, and his behavior was affected.

Many other examples of this process could be cited. For example, the continuing controversy over UFOs show some similarities. There are persons, including a few scientists, who are convinced of the reality of unexplained physical phenomena, but who can't stand the uncertainty of its significance, and become preoccupied with the idea of intelligent visitors from outer space in a manner very reminiscent of Lowell. Others, also troubled by the apparent implications of UFO observations, tend to deny the reality of the observations.

They may be just as incorrect in their appraisal as the "Little Green Men" adherents. The truth, in fact, may be that we must live with and acknowledge this uncertainty for some time. As stated earlier, discovery ultimately takes place in the minds of men. Such minds must be able to carry along many alternative explanations of diverse phenomena until really decisive information becomes available.

The controversy created by Lowell obscured the real questions about Mars, particularly its supposed similarity to the Earth. In rejecting intelligent beings on Mars, scientific opinion tended to accept uncritically the likelihood of abundant plant life to explain the seasonal markings. Indeed, this legacy bedeviled one highly skilled observer as recently as the late 1950s, when he found faint, but real, absorp-

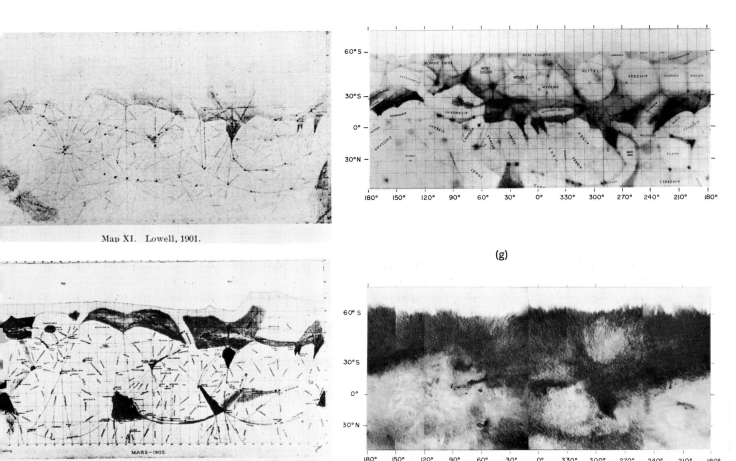

Map XI. Lowell, 1901.

Map XII. Lowell, 1905.

(f)

(g)

(h)

tion features in Mars spectra in the region around 3.5 microns wavelength.

Because Martian plant life had become part of the accepted scientific opinion, he pointed out that many kinds of plant life are known to exhibit features at about this wavelength interval. Later other scientists recognized that the faint features correspond better to the spectrum of HDO, related to the "heavy water" of atomic bomb notoriety. Yet, again geocentrism entered the picture. Instead of realizing that the substance observed was in the Earth's atmosphere and hadn't been correctly removed from the Mars spectra, they postulated that water on Mars somehow had become enriched in HDO over normal water a thousandfold! Such a result could only have come about if large quantities of normal water once had been present on the surface of Mars over long periods of time—that is, under very Earth-like conditions.

The persistence of Lowell's Legacy is truly remarkable, even pervading the interpretation of some of the results from Mariner 7 in August 1969. In that case, one of the investigators initially confused the spectral signature of solid dry ice found in the reflected infrared radiation from the polar cap with what might have been expected from small amounts of the gases ammonia and methane in the atmosphere over the cap. Thus he was led to postulate (and later withdraw) the notion that an exciting clue to possible biological activity had been found. Lowell's Legacy prevented that man from recognizing instead the first diagnostic evidence that the polar caps indeed are

61

Table 4.2   Comparison of U.S. and Soviet planetary missions

composed of solid $CO_2$ and not $H_2O$; thus the caps are colder and even more hostile to hypothetical life forms than had been previously imagined.

There is in fact not a single Mars observation demonstrating organic constituents, much less biological ones. Furthermore, the very low nitrogen and carbon dioxide abundances compared with the Earth and Venus, the absence of a magnetic field and probably of any appreciable core, and the survival of densely cratered terrains perhaps since planetary formation—all suggest that Mars may never have possessed a dense atmosphere like the Earth and Venus and that there probably never was a hypothetical favorable period when life could have begun. Mariner 4 relocated Mars in cosmic genealogy from little brother to at most cousin of the Earth— and apparently to closer kinship with the Moon. And Mariners 6 and 7 have established an independent identity for the "Red planet." Certainly the a priori probability of finding life there has been diminished somewhat in this process of relocation.

However, basic scientific guidelines for U.S. planetary exploration were laid down with authority *before* Mariner 4 returned the information which changed the image of Mars. The search for life on Mars was a popular theme from the beginning of the U.S. civilian space program and was given the status of doctrine by the Space Science Board of the National Academy of Sciences in October 1964:

The primary goal of the national space program in the exploration of the planets is Mars: it is one of the nearer planets (and hence relatively accessible); as a planet, its biological, physical, chemical, geophysical, and geological properties are at least as interesting as those of any of the other planets; of even greater significance and excitement to mankind, it affords the more likely prospect of bearing life.

Subsequent studies sponsored by the SSB have argued for higher priority for the other planets, a point reiterated by the President's Science Advisory Committee in 1967. However, the Mars orientation of U.S. planning has continued without deviation to the present despite repeated funding rebuffs by Congress and growing Soviet competition. The strong attempts to carry out a big Mars program led to passing up relatively inexpensive opportunities for a "first look" at Mercury in 1970, a significant capsule or balloon mission to Venus in 1972, and even a simple Mars lander in 1969 or 1971. Thus it can be argued that "Lowell's Legacy" continues to bedevil U.S. planetary exploration because we have not utilized what proved to be very limited resources in the most opportune way to learn about the planets in general—indeed not even about Mars itself considering the failure to proceed with a preliminary "cheap" lander for the 1969 or 1971 opportunity. Ironically, the first landing on Mars may well be a Soviet one in 1971 or 1973, a possibility apparently not seriously considered in the scientific planning

| | U.S. | | Launch window | Soviet | | |
|---|---|---|---|---|---|---|
| Mission | Weight | Name and date | | Name and date | Weight | Mission |
| | | No launch | 1960 **Mars** | _____, 10 Oct. | | F |
| | | | | _____, 14 Oct. | | F |
| | | No launch | 1961 **Venus** | _____, 4 Feb. | | F |
| | | | | Venera 1, 12 Feb. | 645 kg | Impact (f) |
| F | | _____, 22 Jul. | 1962 **Venus** | _____, 25 Aug. | | F |
| Flyby | 203 kg | Mariner 2, 26 Aug. | | _____, 1 Sep. | | F |
| | | | | _____, 12 Sep. | | F |
| | | No launch | 1962 **Mars** | _____, 24 Oct. | | F |
| | | | | Mars 1, 1 Nov. | 895 kg | Photo flyby (f) |
| | | | | _____, 4 Nov. | | F |
| | | No launch | 1964 **Venus** | _____, 27 Mar. | | F |
| | | | | Zond 1, 2 Apr. | | Entry capsule (f) |
| F | | _____, 5 Nov. | 1964 **Mars** | Zond 2, 12 Nov. | | Entry capsule (f) |
| Photo flyby | 261 kg | Mariner 4, 28 Nov. | | Zond 3*, 18 Jul. '65 | | Photo flyby (f) |
| | | No launch | 1965 **Venus** | Venera 2, 12 Nov. | 963 kg | Photo flyby (f) |
| | | | | Venera 3, 16 Nov. | 963 kg | Entry capsule (f) |
| | | | | _____, 23 Nov. | | F |
| | | No launch | 1967 **Mars** | No launch | | |
| Flyby | 245 kg | Mariner 5, 14 Jun. | 1967 **Venus** | Venera 4, 12 Jun. | 1108 kg | Entry capsule |
| | | | | _____, 17 Jun. | | F |
| | | No launch | 1969 **Venus** | Venera 5, 5 Jan. | 1132 kg | Entry capsule |
| | | | | Venera 6, 10 Jan. | 1132 kg | Entry capsule |
| Photo flyby | 414 kg | Mariner 6, 24 Feb. | 1969 **Mars** | _____, 27 Mar. | | F |
| Photo flyby | 414 kg | Mariner 7, 27 Mar. | | | | |
| | | No launch | 1970 **Venus** | Venera 7, 17 Aug. | 1182 kg | Lander capsule |
| | | | | _____, 22 Aug. | | F |
| F | | _____, 8 May | 1971 **Mars** | _____, 11 May | | F |
| Photo orbiter | 1000 kg | Mariner 9, 30 May | | Mars 2, 19 May | 4650 kg | |
| | | | | Mars 3, 28 May | 4650 kg | |

F Denotes failure to leave earth orbit.
\* Lunar test of planetary spacecraft.
(f) Denotes failure to achieve planetary objectives.

of the U.S. program. The outstanding U.S. contribution to Mars exploration has been large quantities of high quality flyby photography and other remote measurements in 1969; even vaster quantities are expected from an orbiter mission in 1971. An ambitious lander and supporting orbiter, called Viking, is presently scheduled for 1975.

## 4.3 SOVIET PLANETARY SPACECRAFT AND MISSIONS

Very little information is available on the stated objectives, priorities, and future plans for planetary exploration by the Soviet Union. Hence, the program must be judged on the basis of information on flights, both successful and unsuccessful, carried out so far. In the following, we present a summary description of the Soviet spacecraft and missions to date.

Twenty-seven planetary probes have been attempted by the Soviets since their program was started in 1960 (see Table 4.2). No Venus window has been missed and only one Mars window (January 1967) passed without an attempted launch. However, only Venera 4, 5, 6, and 7 were complete successes. Zond 3, which photographed the Moon outward bound, was not launched during a Mars window and lost communication before Martian distance was achieved.

The Soviets have used three basic spacecraft designs. The first was described only once,

63

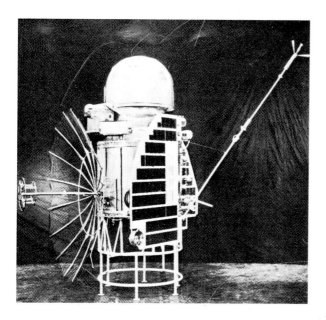

Venera 1 was launched on February 12, 1961, toward Venus, apparently intended as a high-risk impact mission. It failed before reaching Venus but did accomplish the deepest penetration into space at that time. Note that there was no midcourse correction capability.

**4.4 First Soviet interplanetary spacecraft**

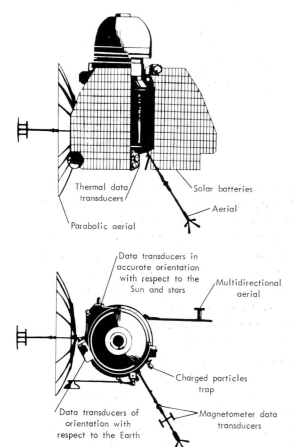

Thermal data transducers

Parabolic aerial

Solar batteries

Aerial

Data transducers in accurate orientation with respect to the Sun and stars

Multidirectional aerial

Charged particles trap

Data transducers of orientation with respect to the Earth

Magnetometer data transducers

under the name Venera 1 (see Fig. 4.4), and consisted of a single cylinder pressurized to approximately one atmosphere. Room temperature was maintained internally by means of painted shutters installed on the cylindrical part of the spacecraft. Solar battery panels and telescoping antennas unfolded following separation of the spacecraft from the booster rocket. The large two-meter parabolic antenna was intended to unfold as the spacecraft approached the planet. The programmer, transmitters, recorders, and the thermal control system were contained in the pressurized body.

On the trajectory to the planet, instrumentation measured the magnetic field, charged particles intensity, and micrometeor counts. The spacecraft carried pendants and the coat-of-arms of the USSR, as is customary on Soviet impactors. Apparently that mission was the same as that of Luna 2, which accomplished the first lunar impact on 13 September, 1959, after making magnetic field measurements.

## 4.5  Mars 1

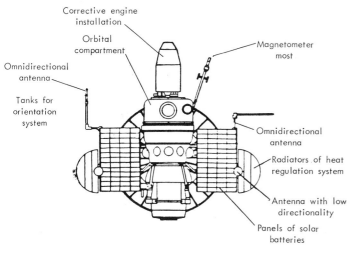

Corrective engine installation
Orbital compartment
Magnetometer most
Omnidirectional antenna
Tanks for orientation system
Omnidirectional antenna
Radiators of heat regulation system
Antenna with low directionality
Panels of solar batteries

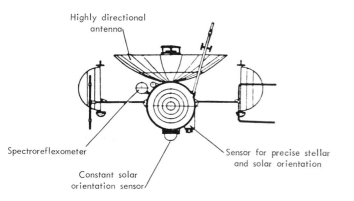

Highly directional antenna
Spectroreflexometer
Constant solar orientation sensor
Sensor for precise stellar and solar orientation

The Soviet Mars flyby attempt was launched November 1, 1962, and failed before reaching Mars. This same basic spacecraft design is evident in all subsequent planetary probes to both Mars and Venus about which the Soviets have released details (see Table 4.2).

The second spacecraft design has been launched and described under a variety of names, such as Mars 1, Zond 1, Zond 2, Zond 3, Venera 2, Venera 3, and Venera 4-7. This design was a great improvement over the Venus 1 spacecraft and contained an auxiliary motor so that midcourse corrections to the trajectory could be made. A star seeker (Canopus) was added to the attitude control system, which also contained sun and earth seekers; temperature control was maintained by circulating liquid. A second pressurized vessel was added which contained the experiments dealing with the planet itself. These were of two types; photographic and remote sensing instruments for flyby missions (Mars 1, Zond 3, Venera 2) and a detachable capsule for lander missions (Venera 3, Venera 4, Venera 5, Venera 6, Venera 7, and Zond 2 [?]).

Zond 1 was the first such spacecraft successfully launched toward Venus. Its mission was never announced; however, Zond 1 did perform

65

(a)

(b)

(c)

**4.6 Soviet Venus flyby spacecraft**

(a) Venera 2, a photographic flyby generally similar to Mars 1 which failed only four days before Venus arrival in early 1966. (b) Windows in the hermetically sealed instrument compartment of Venera 2 through which a film/readout camera, ultraviolet spectrometer, and infrared spectrometer presumably were to view the planet. (c) Waveguide from the instrument compartment to the central housekeeping section of the spacecraft presumably to transfer video signal to the communication system.

two trajectory corrections, so it was probably on its intended flight path when communication was lost. Zond 1 probably was intended to deliver a landing capsule and may have actually impacted the planet.

The next spacecraft successfully injected onto interplanetary trajectory toward Venus was Venera 2, shown in Fig. 4.6. This flyby spacecraft carried a camera in the pressurized payload compartment. Presumably, the payload of Venera 2 was similar to that of the Zond 3 which took pictures of the back of the Moon and acquired as well UV and IR spectra. The Zond 3 camera appears to be only a slight modification of the camera used in the Luna 3 spacecraft in 1959, and is described earlier in this book (see section 3.2). Compared to U.S. camera payloads (like that of the Lunar Orbiter) this Soviet system is rather primitive. Venera 2 was on a trajectory that passed within 24,000 km of the surface of Venus; when the final command was given to the spacecraft to enter its automatic picture-taking program, communications were lost. It is not known if the spacecraft actually did

photograph the planet. There was to be an effort to re-establish radio contact with Venera 2 when it next approached Earth, but there has been no subsequent mention in the Soviet literature of this attempt, so presumably the results were negative.

The launching of the companion spacecraft Venera 3 (see Fig. 4.7) followed that of the Venera 2 by four days, and it arrived at the planet two days after Venera 2. This spacecraft carried a detachable payload of about 900 mm diameter which was designed to survive entry into the Venusian atmosphere. By means of an accurate midcourse correction the spacecraft was placed on a collision trajectory with the planet. Since it was stated that the detachable capsule was thoroughly sterilized before launch, but no mention was made regarding sterilization of the spacecraft, it has

Figure 4.7 (a) Venera 3. Note the substitution of the white spherical entry capsule (900 mm diameter) for the instrument compartment of Venera 2. Venera 3 also failed four days before Venus encounter and crashed silently into its atmosphere. (Photo courtesy of M. Deskau, Interavia.)

Figure 4.7 (b) Close-up of Venera 4 showing the new location of thermal radiator in the high-gain antenna instead of at the ends of the solar panels as on Venera 3 and earlier spacecraft.

## 4.7   Soviet Venus entry probe spacecraft

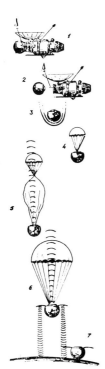

Pictorial presentation of "Venus 4" flight within the planet's atmosphere, descent of the released apparatus and its landing.
(1) Start of the pre-planetary stage
(2) Detachment of the landing apparatus from the orbiting vehicle
(3) Atmospheric braking of the apparatus
(4) Opening of the braking parachute
(5) Opening of the main parachute and the beginning of radio transmission from the landing apparatus
(6) Radioaltimeter begins its functioning
(7) Landing

Figure 4.7 (c) Sequence of delivery of Venera 4 capsule into the Venus atmosphere. This is a copy of a Soviet press release which incorrectly shows the capsule reaching the surface in step 7. In fact, the entry capsules Venera 4, 5, and 6 all were destroyed by the intense pressure well before reaching the actual surface of Venus.

Figure 4.7 (d) The Venera 4 entry capsule, 1000 mm in diameter.

been suggested that the spacecraft would make a trajectory change after releasing the capsule. However, this deflection maneuver was not attempted by the successful Venera 4, so presumably such a maneuver was not intended for the Venera 3 either. Apparently the Soviets assumed that no contaminating microorganisms would survive the atmospheric entry of the spacecraft bus. Communication was lost with Venera 3 just before separation, so both capsule and bus crashed into the planet.

Venera 4 was launched during the next window in June 1967. This mission was successful and justified confidence in the basic spacecraft design after the many disappontments. As a result of modifications in component design, this spacecraft's weight was 1106 kg in contrast to 960 kg for Venera 3. The radiators were moved from the tips of the solar panels to the center of the large antenna. The diameter of the landing capsule was increased from 900 mm to 1000 mm and its weight increased to 383 kg.

Venera 4 was launched June 12, 1967, and on July 29, 1967, the midcourse motor was fired to modify its path and place it on an impact trajectory with the planet. On October 18, 1967, when the spacecraft was 45,000 km from Venus, the large antenna was oriented toward Earth. From this position the capsule automatically separated from the spacecraft upon entry into the atmosphere. The capsule decelerated in the atmosphere from approximately 10,000 m/sec to 300 m/sec, at which

time the braking parachute opened. At an altitude of 52 km the main parachute opened and radio transmission to Earth began (see (Fig. 4-7C). Communication was received for 94 min before termination. Although the Soviets considered originally that this termination corresponded to arrival at the planetary surface, subsequent comparison with Mariner 5 and ground-based radar studies clearly show that the real surface was still 25 kilometers or so away. Analysis of the combined data reveal that the surface pressure is a record 70 atmospheres or more, the surface temperature 800°-900°K, and that the atmosphere is nearly entirely composed of $CO_2$. In May 1969 two more Soviet atmospheric probes denoted Venera 5 and 6 successfully entered the Venusian atmosphere and returned data to Earth before being destroyed by the tremendous pressures there. Finally, in December 1970, the Soviet's persistence was rewarded when Venera 7 survived penetration through the entire Venusian atmosphere, delivered Soviet medallions to the surface itself, and transmitted faintly from there for about 20 minutes.

In summary, then, the Soviet planetary efforts from 1960 to the present reveal the following, consistent pattern:

1) As has been the case with their lunar exploration efforts, the Soviets have placed very great emphasis on an early planetary impact with or without return of significant scientific information. Attempts at planetary exploration started in 1960; the first successful

launching occurred in 1961 with the Venus 1. This first-generation spacecraft evidently could not return pictorial data from Venus, but did make environmental measurements in outer space. Most likely Venera 1 was designed as an impacter analogous to Luna 1 and 2, which carried only magnetometers to impact the Moon.

2) The Soviets have utilized a standardized flyby bus spacecraft for most planetary attempts since 1962. This vehicle, first described in conjunction with the Mars 1 flight, can substitute a small spherical entry capsule in place of the camera compartment.

3) The Soviets evidently have attempted to deploy at least one flyby camera spacecraft and one flyby capsule at each Mars and Venus launch window from 1962 to about 1967. Later launches may have focussed on landers only. The unexpectedly low pressure discovered by Mariner 4 may have made obsolete the landing system the Soviets had intended to utilize for the 1967 Mars opportunity, requiring a whole new approach with consequent delay.

Recently, the Soviets have flown a larger spacecraft system around the Moon and then back to Earth, where film and other portions of the payload were recovered. These flights, denoted Zond 5, 6, 7, and 8, may be precursor to similar flyby and return missions to Mars or Venus in the early and middle 1970s, although the particular spacecraft for the Zond missions was primarily intended for a manned mission. However, components of the basic spacecraft may also be used to transport a survivable lander to Mars in 1971 or 1973.*

## 4.4 U.S. PLANETARY SPACECRAFT AND MISSIONS

The initial U.S. planetary attempt was launched in 1962 toward Venus. Initial plans had called for the Atlas/Centaur launch vehicle; however, this vehicle did not become available until 1966, and the originally conceived 1200-lb spacecraft was reduced to 450 lb in order to be flown on an Atlas/Agena (see Fig. 4.8). The entire process of redesign, fabrication, testing, and launch took only one year. The first launch attempt was unsuccessful due to booster failure, but the second successfully injected Mariner 2 toward Venus on August 27, 1962. Mariner 2 flew within 35,000 km of Venus's surface on December 14, 1962, becoming man's first successful planetary space probe. Particle and field measurements were extended into the orbit of Venus for the first time, revealing that its magnetic field is less than a hundredth that of the Earth. Limited high-resolution infrared and radio emission measurements were also made. Mariner 2's principal accomplishment, however, was the establishment of the lifetime and communications capability necessary for all subsequent planetary probes and, especially, for the epochal Mariner 4 mission to Mars which was carried out two years later.

* The recently launched spacecraft, Mars 2 and 3, represent a third basic design and may attempt this mission.

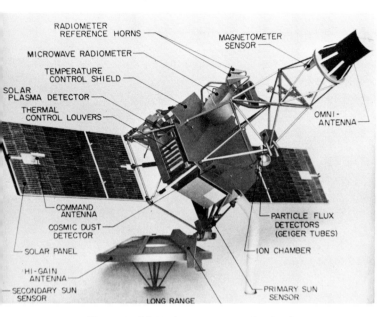

RADIOMETER
REFERENCE HORNS

MAGNETOMETER
SENSOR

MICROWAVE RADIOMETER

TEMPERATURE
CONTROL SHIELD

SOLAR
PLASMA DETECTOR

THERMAL
CONTROL LOUVERS

OMNI-
ANTENNA

COMMAND
ANTENNA

COSMIC DUST
DETECTOR

SOLAR PANEL

HI-GAIN
ANTENNA

SECONDARY SUN
SENSOR

LONG RANGE

PARTICLE FLUX
DETECTORS
(GEIGER TUBES)

ION CHAMBER

PRIMARY SUN
SENSOR

Figure 4.8 (a) Mariner 2 spacecraft with instruments and subsystems identified. Mariner 2, the first successful planetary probe, flew by Venus in 1962 and discovered that there was no magnetic field.

Figure 4.8 (b) Same, in stowed configuration used for launch.

Even before the successful launch of Mariner 2 toward Venus in August, 1962, design studies were proceeding at the Jet Propulsion Laboratory for a flight to Mars in 1964; these studies were based on use of the Atlas/Agena system and as much of the Venus spacecraft hardware as possible. In November, 1962, partly as a result of the launch and initial flight success of Mariner 2, the go-ahead was given for the 1964 Atlas/Agena mission to Mars, to be termed Mariner C.

Mariner C spacecraft to Mars necessarily was limited to about 575 lb, less than half the weight of the Atlas/Centaur spacecraft originally planned for the U.S. program and less than a third the weight of the first Soviet Mars probe Mars 1, which had just been launched by the USSR (see Fig. 4.5). Indeed, there was considerable question whether Mariner C could carry out scientifically meaningful observations at the planet. The carrying

of any kind of landing capsule was out of the question. Weight limitations even ruled out mechanization of the high-gain antenna so that it could be reoriented periodically for tracking the Earth. Ingenious spacecraft design permitted utilization of a fixed high-gain antenna which could be used without reorientation of the spacecraft. This resulted in a simpler operational sequence than that used by the Soviet Mars 1 probe. The mission was severely limited in photographic capability by the over-all weight limitation and by the limited time available for development; the total amount of pictorial data that could be broadcast back to the Earth was only a tiny fraction of the amount that might be considered adequate for viewing an unknown planetary surface. The twenty-one Mariner 4 pictures contain, altogether, about as many picture elements as a single Ranger photograph contains, and only about 1 percent of the elements contained in an ordinary 9- by 9-inch (23- by 23-centimeter) aerial photograph. It was not possible to carry infrared or ultraviolet spectrometers.

On the other hand, the orientation system of the Mariner Mars spacecraft (which "locked onto" the sun and Canopus) was more sophisticated than that of the Mariner Venus spacecraft (which locked onto the Sun and the Earth). Also, the Mars spacecraft was designed to transmit much more data than the Venus spacecraft, over greater distances, and to operate unattended in a complete vacuum, internal and external, for more than twice as long.

On 5 November, 1964, an Atlas/Agena rocket operated satisfactorily in an effort to place Mariner 3 on a Mars trajectory. However, a newly designed lightweight shroud for protecting the spacecraft from aerodynamic heating collapsed during ascent, preventing the solar panels and other spacecraft components from unfolding properly. In a brilliant improvisation, within three weeks the nature of the failure had been diagnosed, the diagnosis had been confirmed by laboratory testing, and a new shroud was designed, fabricated, tested, and installed. Thus, it was possible to launch a second Atlas/Agena during the

1964 launch opportunity. On 28 November, 1964, Mariner 4 was injected on a trajectory toward Mars. On 4 December, a mid-course trajectory correction was performed, causing the spacecraft, seven months later, to pass 6188 miles (9900 kilometers) from the planet's surface.

On 14 July, 1965, Mariner 4 flew by the planet, obtained pictures of resolution up to thirty times the best resolution ever before achieved, determined that the magnetic dipole moment of the planet was less than 1/3000 that of the Earth, determined that there were no radiation and dust belts, and discovered that the atmospheric pressure at the surface was significantly lower than had been indicated by terrestrial measurements.* A brilliant technological improvision had suddenly become a historic scientific achievement.

After the completion of the Mariner 4 mission, the National Aeronautics and Space Administration did not plan any further Mariner-class missions to Venus until the 1970s because maximum priority was given to Mars spacecraft for Saturn-class launch vehicles. The scientific basis for this extreme priority, in marked contrast to the Soviets, evidently was the inference that the high surface temperatures indicated by a variety of

---

* This last achievement was accomplished by purposely flying behind the planet as seen from Earth and observing the effect of the Martian ionosphere and atmosphere on the spacecraft's radio signals. Consideration of this opportunity did not arise until *after* design and fabrication of the spacecraft!

Figure 4.9 (a) Mariner 6 (and 7) spacecraft in configuration used during flight. These two spacecraft flew by Mars in 1969, expanding greatly the tiny "first look" provided by Mariner 4 in 1965. Basically the same spacecraft is to be placed in orbit about Mars in late 1971, and another, with some modifications, is scheduled to be flown by Venus and Mercury in early 1974.

Venus radio emission observations, including those of Mariner 2, reflected an environment considerably more hostile to life forms than that expected in at least some localities on the surface of Mars. This priority was substantially unchanged even by the disclosure by Mariner 4 of a far more hostile and lunar-like surface on Mars than many scientists had supposed.

However, Mars exploration using Saturn-class vehicles proved too costly. In the late fall of 1965, the agency reconsidered the original Atlas/Centaur concept and prepared to launch two advanced Mariner flyby spacecraft to Mars in 1969. Each spacecraft would weigh about 900 lb.

In addition, one spare spacecraft remained from the Mariner 4 mission completed earlier in 1965. The decision was made to modify that spacecraft for launch on an Atlas/Agena for another flyby mission to Venus in 1967. This spacecraft was successfully launched on 14 June 1967 and collected refractivity data on the Venus atmosphere which, when combined with the Venera 4 observations and those collected by ground-based radar, finally solved the riddle of whether Venus really had an unbelievably thick, hot atmosphere or whether some other phenomenon was the

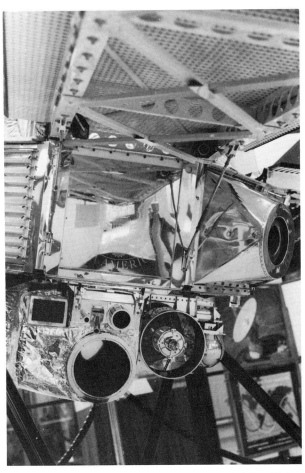

Figure 4.9 (b) Close up of scan platform containing high-resolution TV camera (bottom left), infrared spectrometer to its right and outside of the aluminized thermal blanket. Smaller apertures include an infrared radiometer, an ultraviolet spectrometer, and two infrared limb-sensing devices.

source of the intense radio noise from its atmosphere.

The two flybys to Mars in 1969, Mariners 6 and 7, represented a major evolution in spacecraft design and capability (see Fig. 4.9). Most important, more than a hundred times the data of Mariner 4 was returned, permitting a vast increase in knowledge about Mars.

During 1968 it was decided to outfit two additional spacecraft of the Mariner '69 type with retropropulsion systems and other modifications for use as photographic orbiters of Mars during 1971. A further increase of one-hundred in data return may be achieved with the '71 mission. Thus Mariner 4 was the first close-up peek, Mariners 6 and 7 constituted the first systematic exploration, and the orbiters of 1971 can provide systematic mapping of fixed features and monitoring of variable ones. Truly, the initial period of photographic exploration of that planet will have been completed well before the first U.S. unmanned landing takes place (currently scheduled for 1975), similar in some ways to experience with the Moon.

## REFERENCES

*Icarus*, Vol. 11, No. 1, January 1970, "The Mariner Mars 1971 Experiments."

*Journal of Geophysical Research*, Vol. 76, No. 2, January 10, 1971, "Results from Mariner 6 and 7."

Lowell, Percival, *Mars and its Canals*, 1906.

Murray, Bruce C, and Merton E. Davies, "A Comparison of U.S. and Soviet Efforts to Explore Mars," *Science*, February 25, 1966, Vol. 151, No. 3713, pp. 945–54.

Soviet Space Programs, 1962–65; Goals and Purposes, Achievements, Plans, and International Implications, Staff Report, Committee on Aeronautical and Space Sciences, United States Senate, December 30, 1966.

The Space Program in the Post-Apollo Period, Prepared by the Joint Space Panels, The White House, February, 1967.

# CHAPTER 5
# Mars close-up

## 5.1 CONSTRAINTS AND STRATEGY OF THE PHOTOGRAPHIC EXPLORATION OF MARS

Mars is the only planet other than Earth that has so far been explored by means of close-up photography. After Mariner 4's first look, a pair of ambitious second generation photographic flybys were launched by the U.S. in the early spring of 1969 and returned two-hundred times the photographic data of Mariner 4, including not only a tenfold increase in resolution but also much-needed global photography. In addition, efforts are already underway for a pair of U.S. photographic orbiters in 1971 as well as another pair in 1975 to support the first U.S. lander attempt. In each case, the choices of camera type, coverage/resolution combination, areas to be surveyed, and so on have been, or are being made, in the context of a variety of engineering, fiscal, and historical constraints as well as assessments of the probable scientific value of alternative schemes. Since the experience gained in the case of Mars may provide insight into photographic exploration of other planets and their satellites, and since the decision-making process in this part of the U.S. program can be reconstructed by us, it seems desirable to spend some time on this subject. First we shall review the background of the Mariner 4 photography experiment, and then proceed to the 1969, 1971, and 1975 missions which have evolved in succession from the original Mariner 4 success.

The scientific objectives of the Mariner 4 television experiment were originally formulated in the context of the "Mariner B" project in which it was hoped to send a 1500 lb spacecraft to Mars in 1964. Under these circumstances, approximately $10^8$ bits of photographic data were anticipated to be returned, making possible a high and low resolution survey utilizing two cameras. Such a configuration is most desirable to bridge the "resolution gap" separating Earth-based resolution (best~100 km) from the one kilometer scale, which was initially felt to be the minimum necessary for exploring an unknown planetary surface. The demise of the "Mariner B" project, and the concomitant reduction of 1964 flyby weight to less than 600 lb, imposed very serious restrictions for photography: Only a single, small camera could be carried, only one axis of pointing control would be available, and, most serious, the total photographic data return would be only 1/30 that planned for "Mariner B." *All 21 Mariner 4 photographs together contain but a small percentage of the data in a single high-resolution Lunar Orbiter photograph!* Thus, it is easy to see how the Mariner 4 experiment was designed primarily to investigate the photographic properties of the planet so that some subsequent mission of greater capability could carry out photographic reconnaissance effectively. Of course, scientific discovery was still hoped for, but not required. For example, some color overlap was included, despite the

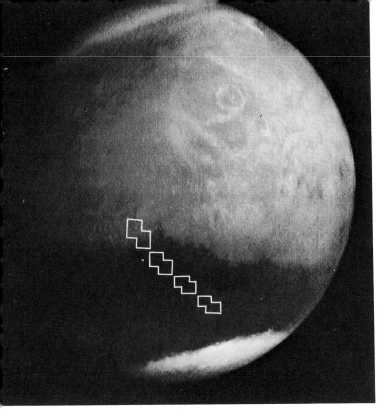

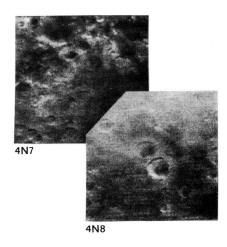

4N7

4N8

**5.1 Mariner 4 frames located on Mariner 7 picture**

Eight of the 21 frames acquired by Mariner 4 on July 15, 1965, are shown in outline on frame 7F76 acquired by Mariner 7 on August 5, 1969. Processed and enhanced versions of the eight frames are reproduced to the right.

paucity of total data, in order to evaluate the visibility of the surface in different colors and to determine if there is spatial variation in surface color below Earth-based resolution.

The choice of ground resolution was particularly difficult with the reduced data return. Degrading the resolution to a value significantly worse than the one kilometer criterion originally envisioned for "Mariner B" raised the real prospect of simply not recording any surface topographical detail: Yet, a one-hundred-fold step in resolution for a few isolated areas on the surface without additional frames of intermediate resolution was likely to lead to uninterpretable photographs, as is discussed more fully in Appendix A.

In addition, other experiments, such as the magnetometer, trapped radiation detectors, and radio-frequency occultation desired as close a pass as practical—yet the planetary quarantine requirement mitigated against trajectories too close. A compromise was chosen resulting in a maximum ground resolution of about 3 km (1.5 km per television line) which permitted a nearly continuous swath of pictures to be acquired (see Fig. 5.1). This step from 100 to 3 km ground resolution was also noted to be about the same as the naked eye versus early telescopic resolution for the Moon, a step that had proven of historic importance because the characteristically cratered lunar surface becomes apparent with that change. Indeed, the Moon remains grossly similar in appearance down to about 0.01 kilometer resolution, as shown in Fig. 3.1. As events turned out, Mars also exhibits a cratered surface similar to that of the Moon, so the particular choice of resolution step proved highly significant there too. However, Mars might have resembled the Sahara Desert, or some other relatively featureless terrain, as many scientists had imagined, in which case there would have been little detail visible at 3 km resolution.

Acquiring the pictures in the "string" of Fig. 5.1 also satisfied the additional objective of viewing the surface over a range of solar illumination angles from near vertical through the terminator. In this way also a limb picture

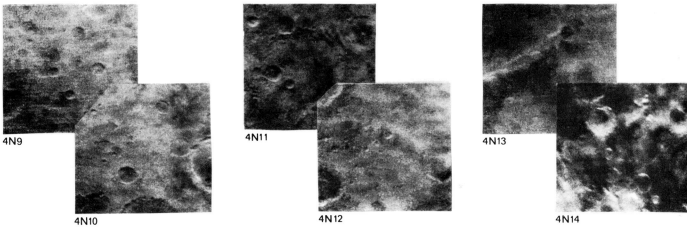
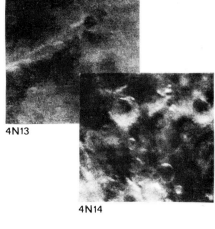

4N9

4N10

4N11

4N12

4N13

4N14

was insured with possible information about the atmosphere. It also should be pointed out, however, that in order to maximize the information stored in the tape recorder, which was the limiting factor of the entire experiment, it was necessary to continuously read picture information into the recorder and to initiate the tape movement coincidentally with the first limb picture. Hence, it would not have been possible to take all the pictures near the terminator in any case. Indeed, the dominance of the tape recorder constraint can hardly be overemphasized. A dramatic increase in transmitter power, spacecraft antenna size, or ground-based receiving equipment performance *would not have increased total data recovered at all.* The data merely could have been returned more quickly! Even if a significantly larger recorder had been flight-proven and available, it would also have needed the capability to *accept* data at a proportionately higher rate.

The choice of track across the planet was influenced by a desire to sample a range of latitudes, particularly near the edge of the polar cap. Yet again, other constraints were involved, including geometric aspects of the radio occulation experiment. Most significant was

the engineering requirement that the flyby take place at a time of optimum viewing from the Goldstone, California, tracking station. The rotation speed of Mars is very nearly the same as that of the Earth. Thus, for the limited range of possible arrival dates at Mars dictated by the need to fly a minimum energy trajectory, the available range of longitudes on Mars was determined by engineering rather than scientific considerations.

Thus the design of the Mariner 4 camera system and mission profile necessarily involved complex trade-offs between what was scientifically desirable and what was possible within the severe engineering, fiscal, and schedule constraints. The resulting camera, whose parameters are summarized in Table C-1 (a), was the first all-digital television camera system as far as we are aware. It weighed only 11 lb, and drew around 8 watts; the associated digital tape recorder weighed only 16 lb.

Perhaps the most outstanding property of the camera system was the remarkable signal-to-noise achieved by the vidicon tube. The limit to intensity discrimination under full illumination was imposed by the limited gray levels associated with 6-bit encoding, rather than target or pre-amp noise. Because of this

77

unusually precise discriminability, and because of the digital form of the data, elaborate processing by computer was possible on the ground after the flight. As can be seen from Fig. A.1, even the best unenhanced pictures proved to be barely useful, whereas the processed versions permitted considerable scientific interpretation. Had the designers of the Mariner 4 experiment chosen instead to trade off intensity discriminability by using, say, only 3 bits per pixel instead of 6, in order to acquire twice as many pictures on the same tape recorder, the results would have been virtually no useful photographs!

Thus the engineering conservatism introduced into the Mariner 4 television design in terms of what could have been deemed excessive intensity discriminability proved to be of overriding importance because the photographic properties of the surface are less favorable than anticipated, that is, there is little scene contrast. Additionally, the camera suffered optical degradation from some unknown source; this phenomenon further reduced the apparent scene contrast. The major risk consciously taken in the experiment was the step of 30 in resolution with only about 1 percent of the surface covered, barely contiguously.

The Mariner 4 television experiment could have successfully determined the photographic properties of a wide variety of hypothetical Martian surfaces, as was its intention. The fact that the experiment also proved to be of historic scientific significance depended on the good fortune that the surface could be explored with less than $5 \times 10^6$ bits and because the tiny amount of data were acquired in a manner which did not preclude discovery because of a priori misconceptions. The implications for "first looks" at Mercury, Venus, Jupiter and its satellites, and the outer planets, are obvious. Beware of a priori ignorance! Or, in the words of Pasteur:

"Preconceived ideas are like searchlights which illuminate the path of the experimenter and serve him as a guide to interrogate nature. They become a danger only if he transforms them into fixed ideas—that is why I should like to see these profound words inscribed on the threshold of all the temples of science: "The greatest derangement of the mind is to believe in something because one wishes it to be so.""*

The Mariner 6 and 7 photographic flybys were designed under entirely different technical constraints to meet new scientific priorities which reflected the results of the Mariner 4 mission. The resultant photographic system, whose parameters are also listed in Table C.1 (a), was not only considerably heavier, but far more complex than that of Mariner 4.

Unlike Mariner 4, the Mariner '69 mission was not primarily weight limited. The Atlas/ Centaur rocket was capable of projecting about 1500 lb to Mars in 1969; yet the actual spacecraft weighed only about 900 lb. Fiscal

* Pasteur, L., 1854. Reported in Dubos, R. J., 1964, *Louis Pasteur* (London: Gollancz).

and schedule constraints associated with the "last minute" decision to fly a U.S. Mars mission in 1969 required that the basic Mariner 4 type of spacecraft structure be flown. Nevertheless, inventiveness and aggressiveness resulted in a great increase in performance. For example, about 100 times the picture data of Mariner 4 was received from Mariners 6 and 7.

The Mariner '69 spacecraft was severely volume limited, and, to a lesser extent, money-limited. On this basis, two Mariner 4 type tape recorders were all that could be flown even though three new planetary instruments were added to the payload—infrared and ultraviolet spectrometers and a 2-channel infrared radiometer. The Announcement of Flight Opportunity indicated a total of $10^7$ bits storage would be available; perhaps only half of that could have been made available to the television systems. Thus the photographic data return from each of the proposed 1969 flybys appeared destined to represent little improvement over Mariner 4. This would not have permitted any increase in ground resolution nor even a systematic extension of Mariner-4-type photography to more than another few percent of the unexplored Martian surface.

Inasmuch as the volume limitation of the spacecraft restricted the weight to around 900 lb, the extra launch vehicle thrust could then be used to fly a faster (and higher energy) trajectory. In this way, the communication distance was substantially reduced. Also, the new 210 ft dish at Goldstone, California, was available, resulting in an increase by a factor of 6 in data rate received there. As a result of these and various other changes, an increase in nominal data rate by a factor of at least 32 that of Mariner 4 was available whereas the storage was to be increased only a factor of two! In actual fact, the data rate was increased a factor of 2000 due to ingenious and aggressive engineering developments.

In order to overcome the serious mismatch between data storage and data transmission capabilities, and yet still utilize two Mariner 4-type tape recorders, an ingenious plan was devised. One of the two tape recorders was converted to an analog system, but without basic change to the mechanical configuration. Then special electronic processing was designed to magnify and store the low contrast features of the pictures on this recorder. After the planetary encounter, the analog recorder was played slowly into an A/D converter, thence into the other digital recorder, and, thence, transmitted to Earth as digital data. In this manner, the useful photographic data return from each spacecraft was increased nearly a factor of 50 over Mariner 4. In addition, a comparable amount of low-resolution mapping photography systematically covering almost the entire planetary surface was also made possible by acquiring photographs with the high resolution camera and then replaying the tape recorders before encounter. All

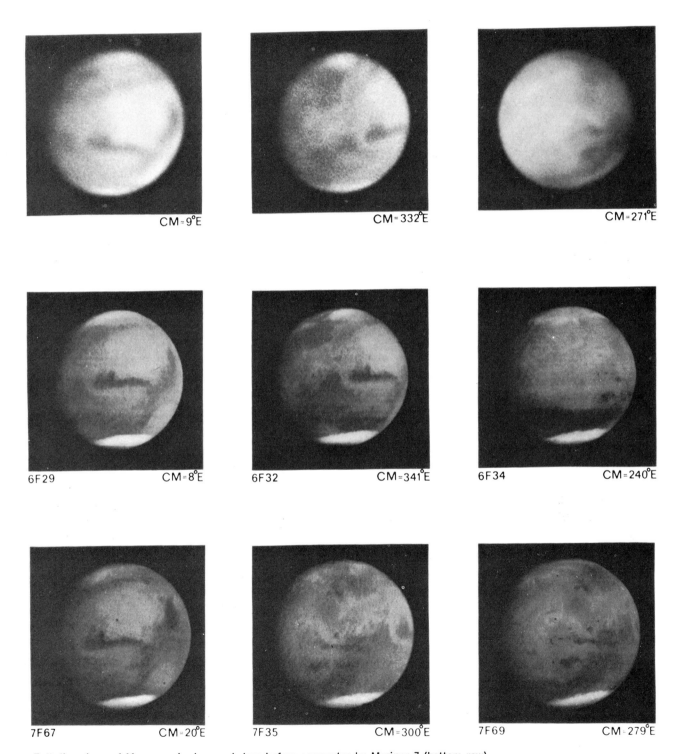

Full disc views of Mars acquired several days before encounter by Mariner 7 (bottom row) and Mariner 6 (middle row) are compared with high quality telescopic views acquired at about the same time by New Mexico State University Observatory.

together, Mariners 6 and 7 returned about 200 times the picture data of Mariner 4 as well as an amount of spectral information about equal to the total picture data of Mariner 4. This whole approach depended on the a priori knowledge gained from the Mariner 4 experiment, that the surface of Mars was much less contrasty and shadowed than that of the Moon.

However this dramatic increase in data storage and in data return (and dramatic reduction in cost/bit as shown in Fig. 1.1) was realized at a cost of more than just dollars: (1) Substantially increased complexity both in the spacecraft and in the ground reconstruction process; and (2) a significant bias in the camera design to emphasize recovery of very low contrast detail. Mariners 6 and 7 were, by this means, endowed with the capability for great discovery. Indeed these spacecraft provided the first *systematic exploration* of the planet in that major elements were sampled including light and dark areas, a "bright desert," the polar cap, and others. They also provided low-resolution coverage of the entire planet (by use of the high-resolution camera before encounter), Mariner 4 resolution along carefully selected paths for about 20 percent of the planet, and nested high-resolution samples down to several-hundred-meters ground resolution. Altogether, a range of nearly 1000 in resolution was obtained. Figures 5.2 through 5.6 illustrate some of the principal discoveries.

On the other hand, it cannot be ignored that the data system improvisation responsible for the dramatic increase in total data of the 1969 mission was an improvisation—not a logical step in an orderly technological development. Not only is the tape system used inferior to what could have been done with a suitably-designed tape system, but alternative new developments like dielectric tape storage could not have been seriously considered. This had important implications for the 1971 mission.

The ground rules for the design of the U.S. 1971 Mars orbiter mission were that *Mariner '69 hardware be utilized with minimum changes!* Thus the cameras and data systems intended for a few cycles of operation during flyby must be operated for hundreds of cycles during months of orbital operations. Dielectric tape or film cameras would have an obvious application to such a mission. Yet, because of the historical, schedule, and fiscal constraints surrounding the initiation of the latter two Mariner Mars missions, possible new developments in camera/storage technology could not be considered for the Mariner '71 orbiter mission. However, a better version of the magnetic tape recorder was finally adopted, and some improvements to the vidicon cameras have been possible as well.

The 1971 Mars Orbiter mission should permit the return of more than 100 times the picture data of the 1969 mission or a factor of more than $10^4$ over Mariner 4. Thus, there will be the opportunity for systematic mapping

81

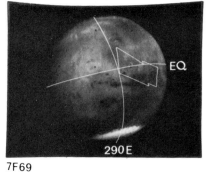

7F69

These Mariner 6 frames provided man's first glimpse of a terrain on Mars unlike any known on either Moon or Earth. The high-resolution frames are about 100 kilometers in maximum dimension, and the low-resolution frames ten times larger.

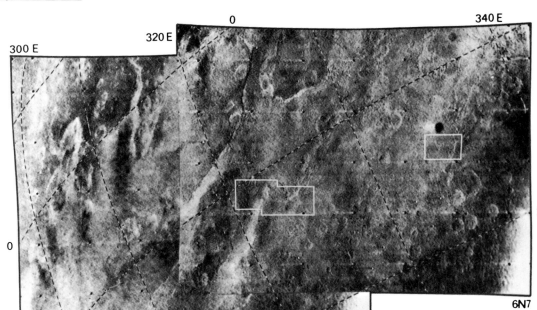

of fixed features and the monitoring of variable features. An increase in ground resolution will result from location of the spacecraft orbit somewhat closer to the surface than was the case with Mariners 6 and 7. At present, about 850 to 1250 kilometers is considered as the closest acceptable distance.

A further difficulty was that most of the redesign of the '69 hardware for the '71 mission had to be carried out before the scientific and performance results of the '69 mission were available. Yet the '71 mission had to be carried in a manner permitting the incorporation of essential information from Mariners 6 and 7 on optimum lighting and on

any latitudinal dependences, on relative visibility with different filters, and on the appearance and possible location of new, unexpected features. Fortunately, the initial results of Mariners 6 and 7 did not indicate that major changes in hardware were necessary for the '71 mission. However, some scientific priorities and nominal mission profiles have warranted reconsideration in view of the variety of terrains, the paucity of clouds, and the diversity of transitory features observed at the edge of and on the polar cap.

Considerations of orbits and spacecraft operations indicate that it is desirable to use one of the spacecraft in 1971 primarily for mapping

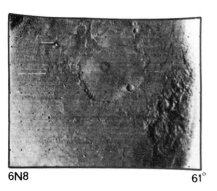
6N8                                        61°

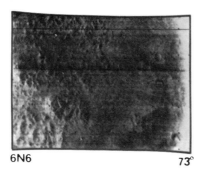
6N6                                        73°

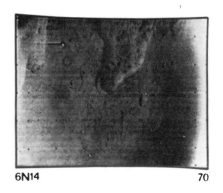
6N14                                       70

of fixed features and the other primarily to monitor variable features. For the mapping of fixed topographic features, systematic photography of successive portions of the Martian surface from about the same altitude and viewing aspect, and especially at the same low-lighting angle, is essential. Study of the wave of darkening, local frost deposits, the polar cap, and cloud phenomena, on the other hand, require frequent return to the *same* area, at the expense of total coverage. In addition, higher lighting angles may be desirable in order to minimize topographic and maximize albedo contributions to the observed terrain brightness. If both spacecraft are successful, then our knowledge of the surface of Mars will certainly increase dramatically once again. Additionally, the Soviets may also orbit Mars in 1971 (or even land), leading to the first simultaneous exploratory venture in outer space. Perhaps the day is coming when there may even be a joint U.S./USSR exploratory venture.

Another pair of Mars orbiters is planned by the United States for 1975 (the sixth and seventh American remote-sensing spacecraft to that planet). They are intended to support two soft-landers labeled "Viking" to be simultaneously deployed on the surface. From lunar experience, a ground resolution of a meter or so is necessary to actually detect the capsule on the surface; a ground resolution of at least 5 meters is probably necessary as well in order to characterize the local environment of the

capsule and to compare that environment with other localities on Mars. Furthermore, such high surface resolution requires a large format, that is, many picture elements per frame. Neither the resolution nor the format are easily achieved with vidicon systems. A film/readout system would be a likely choice; indeed, a similar system was flown around the Moon for precisely these reasons to support the surveyor soft-lander and to help select sites for the Apollo landings. However, the 1975 Viking orbiter has been designed with an entirely new vidicon/tape recorder system of a somewhat higher performance level than that used on the previous Mars flights. An increase in surface resolution of about two and more contiguous coverage are intended. It was felt that use of the film/readout system would be costly and require some new developments. The use of the new vidicon/tape recorder system will, however, also cost significantly more than previous Mars imaging systems and also require new developments. It is to be hoped that the long history of initially overconstrained Mars missions with dramatically rewarding results will extend to the Viking mission also.

The results from the Mars photography of the 1970s will be enthusiastically received by scientists and laymen alike since each step will represent some kind of advance over previous efforts. Nevertheless, it is still proper to ask: Does each step represent the efficient and thoughtful use of available technological

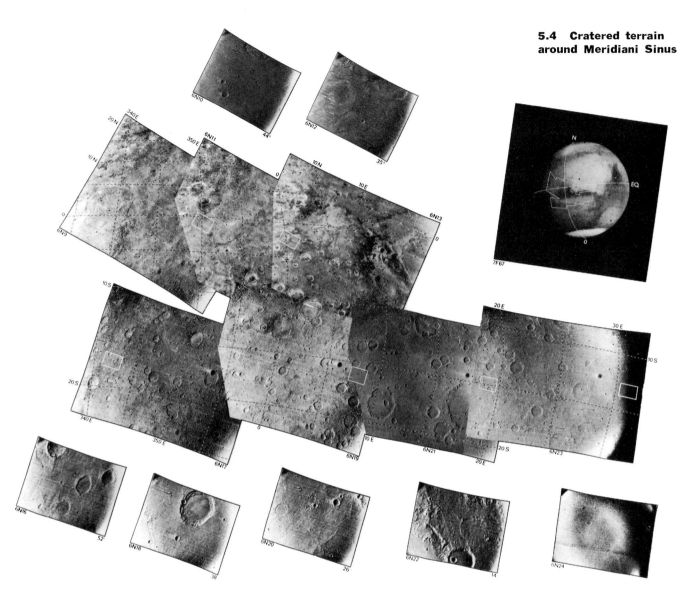

An area comparable in size to most of the United States is included in this mosaic of Mariner 6 frames. Craters are seen everywhere, but with a distinctive difference in form between the larger (more degraded) and smaller (fresher) ones.

resources to accomplish its purposes in the most economical or most productive way possible? As the exploration of a planet proceeds by successive steps toward more specific experimentation and that object becomes a more familiar place, then the need to demonstrate such cost/effectiveness becomes more compelling.

As we will see in Chapter 7, the Mars program of the United States for the 1970s could unintentionally lead to a rather inefficient and unbalanced use of very limited resources for planetary exploration. In particular, earlier proposals for Mars flights at every opposition ('69, '71, '73, '75) simply could not incorporate efficiently preceding results. Other valuable and obviously cost-effective mission opportunities have been and continue to be

**5.5 Cratered terrain at high southerly latitudes on Mars**

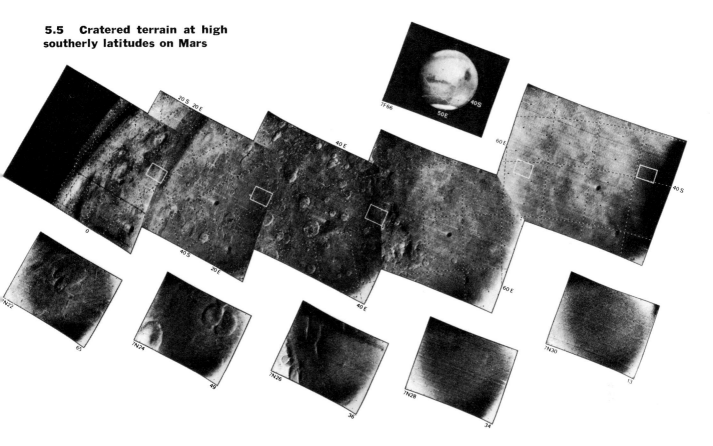

Craters are seen over a broad reach of the southern hemisphere of Mars in this mosaic of Mariner 7 frames except in the bright circular "desert" known as Hellas. The featureless topography there remains an enigma.

passed up in order that a proposed "crash" effort on Mars continue; meanwhile, the "crash" effort itself cannot afford to be cost effective. The Mars program has been proposed repeatedly at a rate inconsistent with available resources and total exploratory opportunities; twice now, at the beginning of 1966 and of 1970, the Congress and the Bureau of the Budget have put the brakes on overambitious plans, and twice now dedicated scientists have seen their planning and design efforts wasted. More modest Mars programs in both cases probably would have been supported and successfully executed. The value of planetary exploration to the United States is as genuine scientific achievement, reflecting good collective judgment and balanced perspective. We have the opportunity to demonstrate as a nation to ourselves and to the literate world those qualities of imagination and style which are so admired in outstanding individuals.

## 5.2 THE MARINER PICTURES AND THEIR SCIENTIFIC SIGNIFICANCE

Perhaps the most important single justification for space photography is its exploratory potential. More than any other kind of observation, a picture can reveal an aspect or condition of the planet which was not even imagined beforehand. The photographic exploration of Mars certainly can be regarded as evidence for this view. From Earth-based photography to the Mariner 4 experience and on to the Mariner 6 and 7 results of 1969, many of the principal results were never

**5.6    Mars and Moon compared—I**

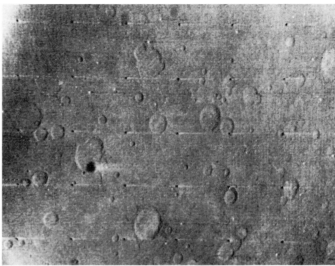

The far left picture of the lunar uplands has been reduced in contrast to match that of the three accompanying Mariner 6 wide-angle frames. The greater vertical relief is evident from the wider wall projections.

specifically considered by the experimenters beforehand. On the other hand, a picture is not limited in its usefulness simply to the unexpected things it might reveal. A surprisingly long list of important specific scientific objectives of Mars photography can be formulated. Thus any particular mission must develop a strategy for pursuing such objectives without compromising the capability for exploratory discovery. In the case of Mars, both fixed and variable features exist and scientific objectives are easily grouped into corresponding categories.

The tiny glimpse provided by the Mariner 4 pictures could not contribute to an understanding of transient features; however, it did provide unexpected information concerning the existence of cratered terrains on the planet. Mariners 6 and 7 not only extended our knowledge of Martian cratered terrains in resolution and geographic extent, but also discovered at least two other totally unexpected terrains. In addition, the 1969 probes contributed to the understanding of transient features and phenomena such as the polar

caps, seasonal variations, and the "blue haze." The two Mariner orbiters being readied for the 1971 mission will provide the opportunity for a thorough survey of the transient features as well as complete mapping of the major physiographic features.

There are three principal aspects of the *fixed* Martian features for photographic investigation: (1) tectonic classification; (2) crater statistics and morphology; and (3) local surface environments.

*Tectonics*—the study of patterns and processes of deformation—affords one approach to the outstanding geological question concerning Mars raised first by Mariner 4: To what extent is Mars a close relative of the Moon in terms of internal structure and, especially, chemical history? Did Mars ever possess an extensive atmosphere in which the initial development of life could have taken place, or has it always resembled the Moon as much as it does now?

Mountain-building deformation is generally regarded as a sensitive indicator of chemical differentiation of a planetary scale. Hence,

 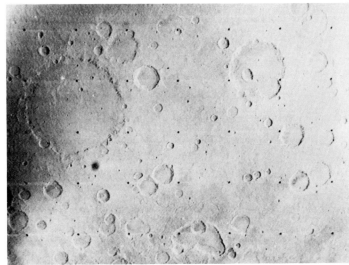

the existence on large-scale topographic features *of internal origin* is a most important surface clue as to internal development. High priority therefore has been given in the photographic exploration of Mars to the search for remnants of arcuate mountain systems, folded mountains, and, generally, endogenic features *unlike* those of the Moon.

As indicated first by Mariner 4, the surface of Mars is generally cratered like that of the Moon but is much smoother. Mariners 6 and 7 confirmed this general appearance for other areas of the planet and also discovered that the bright "desert" Hellas is an extraordinary featureless plain. There are other "nonlunar" terrains, implying the existence of geographic variations in surface processes and/or materials unknown elsewhere in the Solar System. Extension of such photography in 1971 over a large fraction of the planet, and down to about 1-km resolution, as well as selected high-resolution sampling down to better than 100 meter resolution, should permit mapping of what are evidently a considerable variety of morphological provinces on Mars, and thus

provide the basis for unravelling the major elements of Mars history and surface processes. The question of the weathering, transportation, and sedimentational processes responsible for the greater smoothness of Mars compared with the Moon is being attacked with the Mariners 6 and 7 photography by intercomparison of areas on the planet at different latitudes and with different geomorphological associations, as well as analysis of the variety of detail on and about the south polar cap.

Perhaps the most important practical question regarding the search for life on Mars is, where to land? Most specialists expect that any possible life will be highly concentrated in favorable surface environments, perhaps constituting only a tiny fraction of the over-all planetary surface area. Mars exhibits a somewhat broader range of local environments than does the Moon, but the range must be quite restricted compared with that of the Earth. Thus the search, at high resolution, for characteristic differences between local areas, and the inferential identification of distinct classes of local surface environments, is of

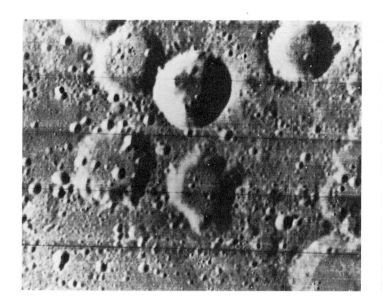

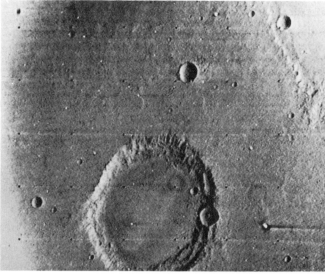

**5.7 Mars and Moon compared—II**

A similar comparison to that of Figure 5.6 except the narrow-angle Mariner 6 frames are compared with a lunar picture of similar scale on the far left. The paucity of small craters on Mars is evident, suggesting an episodic history there of small crater formation and removal.

high priority. This same procedure is also a most desirable way to investigate the question of the nature of the erosional processes on the planet.

The variable features of Mars can be grouped into four categories for consideration of the role of photography in their study: (1) clouds and transient frost deposits; (2) the polar caps and other long-duration frosts; (3) the so-called "blue haze" and other obscurations; and (4) the wave of darkening.

Although the first analyses of the Mariners 6 and 7 close-up pictures failed to yield a single positively identified cloud, the potential significance of television monitoring of Martian clouds and frost deposits from an orbiter, in contrast to a flyby, is very great. Even occasional photographic observation of clouds, for example, can indicate much about atmospheric structure in the cloudy region: the processes forming the cloud, such as radiative cooling, orographic lifting, or thermal convection; the thermal structure of the atmosphere (for example, stable or unstable); and, of great significance, the local atmospheric

moisture content. Such observations would be of great importance in indicating the characteristics of local or temporary surface environments. If "clouds" are comparatively frequent, as may be the case in the "W" cloud area observed in the far-encounter phase of the Mariner 6 and 7 photography, the extended observational capability of an orbiter immediately provides advantages. It might then be possible to determine characteristics of the moisture climatology from cloud and haze distributions. Cloud cover could probably be related to small or large scale topographic features. Perhaps most significantly, wind patterns and their time variations could be mapped across portions of the planet. One special significance of transient frost deposits is that they may provide clues to localities where water vapor is passing into or out of the soil and, hence, to the possible temporary existence of soil moisture.

Depending on orbital and camera mechanization choices, changes in shape of the edge and interior of the polar cap can be observed in 1971, a subject of special interest after the

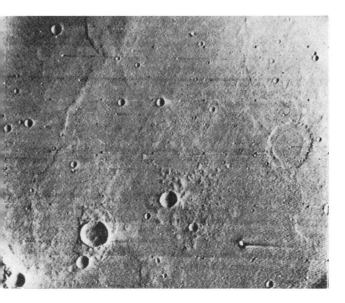

"first look" provided by Mariner 7. Additional data on slopes of the underlying ground may be available from photometric analysis of the uncovered-terrain photographs. And, in any case, it will be most important to view some of those areas completely covered in 1969 by peculiar frost features to learn if such are permanent or transitory.

Ground-based observers had for many years referred to an absorbing "blue haze" on Mars; other workers believed that the phenomenon was a surface effect of some kind. Surface topography is clearly visible in the blue-filtered pictures of Mariners 6 and 7, thus indicating that the blue haze does not exist and that there is indeed an additional strangeness about the planet's surface. Conversely, it can be argued that the risking of the 25 percent of the wide-angle frames taken with a blue filter paid off in the 1969 flights, thus permitting the 1971 mission to operate more effectively. Had there been a blue haze on Mars, these pictures would have shown nothing.

The 1971 mission will provide the first opportunity to study the seasonal variations from the neighborhood of Mars. The 1969 mission has sampled the topography in light and dark areas and may have collected significant clues, especially the albedo markings associated with large craters. Seasonal change in contrast between the bright and dark areas proceeds from the sublimating springtime polar cap toward and past the equator. The borders of the maria appear to become sharper, the maria darken relative to the deserts, and, according to some observers, they may develop blue or green colorations. The effect fades with the advances of the seasons and is followed by a second wave from the opposite pole half a Martian year later.

Seasonal changes in reflectivity of some areas on Mars first suggested to astronomers that Mars might have vegetation. According to this view, the effect is caused by growth of vegetation in response to the seasonal availability of water vapor. Other, nonbiological, causes have also been proposed. These include wind-blown volcanic ash, atmospheric haze, physical changes in the soil due to freezing and thawing, the hydration of hygroscopic salts, and so on. No theory has won general support, and the wave of darkening remains a major mystery of the planet. Photographic observations of interest to be repeated from an orbiter include: the relation of light and dark areas to the local topography, the fine structure of the dark regions, the existence of topographic or other differences between permanently dark areas and those that show large seasonal variations, the question of color changes in the dark regions, and observations on the rate and direction of the "wave" at a higher resolution than has yet been possible. The value of these studies will be enhanced by combining them with measurements of the local temperature and water-vapor concentration.

Another important scientific aspect of space photography is that it provides the basis for high-resolution maps. Topographic maps are needed for both the fixed feature studies such as geology and as a backdrop for the variable feature studies, as the relationship between the topography and surficial change can be very significant. Frequently other remote-sensing experiments depend on accurate maps and sometimes pictures themselves to permit interpretation of their data. The Mariner 6 and 7 photography was carefully designed so that the whole disk photography (far-encounter) could be combined with close-up photography (near-encounter) to produce useful maps. On the other hand, Mariner television cameras, as well as the Lunar Orbiter film systems, were not designed with mapping as a major objective. Thus there are problems of geometrical calibration and difficulties caused by the small format which have hampered efforts to make planetary measurements and obtain photogrammetric results from these pictures. Even so, these techniques have resulted in improved estimates of planetary radii and definition of its geometric shape.

Even such a brief summary emphasizes that space photography of Mars, both past and future, contains enormous scientific potential. Clearly, the design and execution of that photography must constantly reflect interest in and knowledge of its diverse scientific uses if that potential is to be fully realized. The enormous lay interest in close-up photography should not obscure the fact that it constitutes a valid investigation of paramount importance. Photography can be both popular and good science.

## 5.3 PHOTOGRAPHY AND THE SEARCH FOR LIFE ON MARS

The return of new, close-up pictures of Mars is invariably accompanied by a standard question from the press: "What do these pictures tell us about life on Mars, Doctor?" The standard response by the scientists involved is: "One cannot tell anything directly about the existence of life on Mars from images whose best surface resolution is so much greater than the size of possible hypothetical life forms or characteristic surface features." However, the question really does deserve to be asked because the "search for life on Mars" has been used and overused as a means of focusing public support for planetary exploration. And the scientists' answer really avoids the question a bit (for fear of exaggerated public reaction). In fact, the original enthusiasm for possible life on Mars by Lowell and others was totally based upon results from *visual imagery*, that is, observations of the seasonal variations in contrast and the development of polar caps and the remarkable similarity to the Earth in length of day and obliquity. Furthermore, flyby photography has, in fact, played a major role in the development of a more pessimistic scientific assessment starting in 1965 and continuing to the present. Thus photography indeed is related to the search for life on Mars, not just in aiding in the selection of a future landing site where a direct test for life might be made, but also directly in the assessment of probabilities.

Flyby and orbiter photography of Mars is relevant to the possibility of life on Mars because photographic information is pertinent to: (1) the question of whether Mars ever had oceans in which life could have originated in the manner it is supposed to have begun on Earth; (2) the question of how hostile to any possible life is the present surface environment of Mars; and (3) the existence of any indirect evidence of life processes.

Photographic as well as other kinds of data returned by Mariners 4, 6, and 7 have successively decreased the probability of life on Mars by indicating that Mars very likely never had oceans or an aqueous Earth-like atmosphere, and that the surface is much more hostile than previously imagined. Furthermore, no new clues to life processes were uncovered despite the great increase in total data about the planet (see Fig. 1.5).

The television pictures of Mariners 6 and 7 disproved the existence of the "blue haze" and the "polar collar," two interpretations of ground-based imagery which had suggested a less extreme surface environment. In addition, the absence in the Mariner pictures of localized white clouds or frost patches has been a disappointment to those who have hoped the hypothetical local heat sources or other conjectural mechanisms might provide "oases" in what must otherwise be an extremely dry environment; water is not stable in the liquid state at or near the surface due to the extremely low surface pressure.

However, the most telling evidence against life on Mars comes from Mariner pictures showing heavily cratered terrains which resemble the lunar uplands (see Figs. 5.4 and 5.5). Such terrains must have developed after any hypothetical oceanic episode early in Mars' history because we know from terrestrial experience that water erosion (which necessarily must accompany the existence of pools of liquid water on a planetary surface) will

91

erase topography forms within a small fraction of the history of Mars or the Earth. Thus the question of early oceans on Mars boils down to the question of the age of the heavily cratered terrains on Mars. Are they primordial,* like the lunar uplands, or could they have formed later on in Mars' history, still leaving time enough prior for a hypothetical Earth-like era? The question of age of the heavily cratered terrains, in turn, comes down to whether enough asteroids and large meteorites could have impacted Mars in the last 4.5 billion years or so to account for the present surface density of large craters there, say, in the 50-300 kilometer size range? This question in turn can be addressed in terms of the impact flux rates recorded on the Moon, and the probable relation between the flux rate there and on Mars.

With the isotope age dating of the lunar surface localities from Apollo samples, the number per unit surface area of large craters present on, for example, Mare Tranquillitatis, is now known to correspond to all that formed in the last 3.5–4.0 billion years; this surface density must represent half to three quarters of all that have impacted that surface since the Moon attained its final size, that is, since primordial times. This crater density is vastly less than that of portions of the surface of Mars. However, Mars probably has never been subject to a higher impact rate due to its nearness to the asteroid belt, thought to be the source of some or most of the objects impacting both the Moon and Mars. It has been estimated that the flux difference could be as much as 20 to 25 times higher for Mars, although this value probably is extreme since it does not take into account all the possible impacting objects for both bodies. Even with a factor of 25 difference, the Martian surface still turns out to be primordial. Thus it appears quite likely that the large craters of Mars exhibit topography of an age comparable to that of the lunar uplands and likewise are relics of the final phases of planet accretion and formation. If these craters are truly primordial, then oceans could never have existed on Mars and life could not have originated there in the way it is thought to have on Earth.

Life on Mars, however, is by no means ruled out by these arguments; indeed it is virtually impossible to prove the absence of an arbitrary, hypothetical life form on Mars, the Moon, or elsewhere. Life of certain chemical properties assumed a priori can be looked for in a finite number of samples. However, flyby photography has played a major role in relocating Mars on the "cosmic tote board" from an interesting bet to a very long shot. In fact, early oceans cannot be ruled out at present nearly as satisfactorily on Venus or even Mercury as they can on Mars and the Moon.

In summary, then, the notion of life on

---

* Corresponding to the formation of the solid planet generally taken to be about 4.5 billion years ago on geochemical evidence.

Mars developed in the late nineteenth century as a result of ground-based visual imagery and then grew out of proportion for unscientific reasons in the first half of the twentieth century. Since then that notion has been placed into the more appropriate perspective of an extraordinarily exciting but low-probability long shot as a result of close-up photography and other observations. Thus, photography (and visual imagery) indeed have had a very great deal to do with scientific assessment of the likelihood of life on Mars.

**REFERENCES**

*Journal of Geophysical Research*, January 10, 1971, Special Supplement, "The Mariner 6 and 7 Pictures of Mars."

Leighton, R. B., B. C. Murray, R. P. Sharp, J. D. Allen, R. K. Sloan, Mariner IV Pictures of Mars, Technical Report 32–884, Jet Propulsion Laboratory, December 15, 1967.

Leighton, R. B., *et al.*, "Mariner IV Photography of Mars: Initial Results," *Science*, Vol. 149, pp. 627–30, 1965.

Leighton, R. B., N. H. Horowitz, B. C. Murray, R. P. Sharp, A. H. Herriman, A. T. Young, B. A. Smith, M. E. Davies, C. B. Leovy, "Mariner 6 and 7 Television Pictures: Preliminary Analysis," *Science*, Vol. 166, pp. 49–67, 1969.

National Academy of Sciences, "Biology and the Exploration of Mars," Publication 1296, 1966.

Sagan, Carl, *Planetary Exploration*, Oregon State University Press, 1970.

# CHAPTER 6
## Other worlds

### 6.1 VENUS—A WOMAN SCORNED

Venus is the planet most easily observed from the Earth because of its great brightness and relatively frequent appearance, although the very best surface resolution obtainable is not as good as for Mars (see Table 4.1). It was evident from the beginning that the surface was covered with dense clouds because of the high planetary albedo, the absence of conspicuous visible markings or variations in the integral light of the planet, and from the thin crescent of light ringing the planet when seen from the anti-solar direction. Recent information about the very dense atmosphere makes it appear unlikely that the surface is ever visible through the cloud covering. On the other hand, faint markings have been reported by visual observers and more prominent ones photographed in the ultraviolet, as are also shown in Fig. 6.1. The ultraviolet markings are most unusual because their over-all pattern appears to change only slowly from day to day; a definite rotating appearance of these semipermanent features has been reported. Most surprising, the apparent rotation rate of the ultraviolet markings is 4 to 5 days retrograde—very much *faster* than the rotation of the surface, which has been determined with great precision by planetary radar measurements to be 256 days retrograde. The causes and significance of the ultraviolet markings on the planet constitute a principal question about the planet, a question which requires high resolution UV photography from a flyby or orbiter to pursue further. Also, useful information concerning cloud heights and shapes, perhaps bearing on atmospheric structure and dynamics, may be available from close-up visible photography along the terminator and of the limb. Thus Venus would seem to have been a likely choice for at least a Mariner 4 class photographic mission during the 1960s to determine whether photography is a good way to explore the planet and, especially, to address the "first-order" question of the ultraviolet markings.

Surprisingly, although Venus has been the object of six successful space missions out of nineteen attempts (see Table 4.2), no successful photography has been carried out by either U.S. or Soviet space probes. In the closest attempt so far, the Soviet photographic probe Venera 2 (see Fig. 4.6) failed only four days before reaching the planet.

Neither of the two successful U.S. flybys was intended to take pictures, even though the latter, Mariner 5, originally was constructed as a photographic spacecraft, the back-up for the Mariner 4 mission. After the Mariner 4 success, it was decided to retrofit and launch that single spacecraft to Venus. Further, within the very limited costs that had been allotted, it was decided to remove the camera and related equipment in order to make room for a second radio frequency occultation

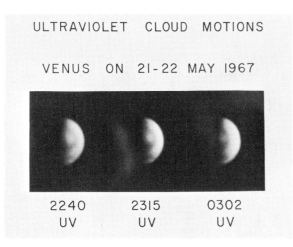

ULTRAVIOLET CLOUD MOTIONS

VENUS ON 21-22 MAY 1967

| 2240 UV | 2315 UV | 0302 UV |

In ultraviolet light, atmospheric markings can be photographed on Venus which appear to rotate with a period of 4 to 5 days—much faster than the 256 day rotation of the solid body. High resolution study of this unexplained phenomenon is a principal objective of the television experiment aboard the 1973 Mariner Venus Mercury mission. New Mexico State University Observatory photographs.

experiment to supplement the original one used so successfully at Mars. Radio probing of the atmosphere was felt to be that much more relevant to the scientific questions about Venus. Neither its exploratory potential or popular significance was sufficient to break loose the necessary additional funds to restore photography to the spacecraft, even though at the same time work was proceeding toward Saturn 5-class exploration of Mars. Indeed the funds expended just on the preliminary design of the Mars Voyager mission, later cancelled, exceeded the total costs of Mariner 5 to Venus.

Late in 1967, after the U.S. Mariner 5 and the Soviet Venera 4 flights, consideration was again given to the eventual retrofitting of what was expected to be a spare Mariner Mars spacecraft for a flight to Mercury via Venus in 1970. This specacraft, the back-up for Mariners 6 and 7, could have returned abundant close-up photography of both planets, including UV coverage of Venus. However, this proposed mission was competitive for

resources with the 1971 Mars orbiter mission mentioned in the last chapter and was passed over altogether.

Since then, approval has been given by the United States for a similar Mercury/Venus photography flyby in 1973, the next dual planet opportunity after 1970 (and the last good gravity assist opportunity to Mercury until the 1980s).

United States planning for such a mission could be complicated by additional Soviet missions to Venus during the 1972 opportunity. Such missions could provide new knowledge requiring changes in experimental strategy. This is but one of many examples of the necessarily *bilateral* character of planetary exploration signalled by the Soviet Venera 4, 5, 6, and 7 successes. In any case, U.S. designers have to consider photography with the same camera system of two very different kinds of planetary targets about whose photographic nature less is known now than was the case for Mars when Mariner 4 was designed.

The U.S. Mercury/Venus mission for 1973 will be a minimum effort, involving only a single launch and stringent control of scientific instrumentation. The equipment used will be controlled largely by that left over from the 1971 Mars missions. It is hoped that improvisation and ingenuity will help unlock the enormous scientific potential. But surely, Venus must wonder at Earth's reluctance to view her contenance!

These composite images of Mercury acquired at New Mexico State University Observatory are among the best in the world—yet show very little detail. Mercury truly awaits exploration.

## 6.2 MYSTERIOUS MERCURY

The inner boundary of the planetary system is represented by Mercury, the only significant object as dense as Earth in the entire Solar System. Mercury surpasses Earth in preponderance of iron and other heavy metals in its chemical makeup. Were it of equal size, Mercury would greatly exceed the Earth's mass and is by far the densest object in the Solar System (see Fig. 6.2). It thus represents an extraordinary planetary body, a unique chemical crucible whose present physical status may provide insight into the nature of planet formation generally, and especially into those processes characteristic of a dense planet like Earth. The degree of development there of phenomena such as mountain-building, formation of an internal core and other kinds of radial differentiation, and the presence of planetary-scale magnetic and electrical fields is of significance to the understanding of how our own planet came to be the way it is. And, the origin of the Solar System itself can be pieced together only from knowledge of the physical configuration and surface histories of the principal objects which compose it. Hence, Mercury would seem to have warranted great interest as an object of study and to represent a prime target for close-up investigation of its surface with automated spacecraft as soon as the requisite technology became available.

Surprisingly, this has not been the case.

Until recently, Mercury has received only cursory attention from optical, infrared, and radio astronomers. Since their opinions have been an important influence in the setting of priorities for U.S. space flights, Mercury occupied a relatively low priority when comparatively inexpensive space flight opportunities first were considered. More recently, Mercury has received stronger endorsement as an exciting target for scientific study. Mars may have been regarded too optimistically over the years as a result of uncritical evaluation of insufficient and sometimes inaccurate observations. Mercury, on the other hand, seems to have been regarded too pessimistically on the basis of uncritical evaluation of an enormously smaller amount of equally inaccurate observations.

Because Mercury never gets more than 28 deg from the Sun, as viewed from the Earth, it must always be viewed in the daytime, or near the horizon during twilight. Such conditions greatly complicate all optical and

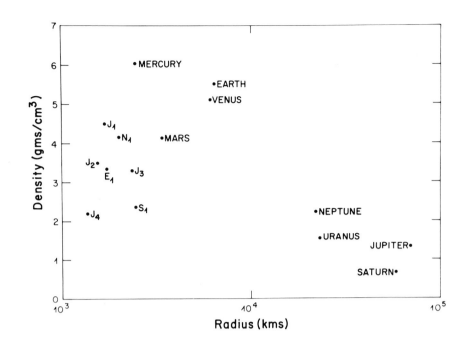

The giant planets are seen to also exhibit uniformly low density; the lunar-sized objects to represent quite a range of density; Earth and Venus are twins; and Mercury is anomalously dense and therefore intrinsically different from all other objects in the Solar System.

infrared observations and, to a lesser extent, radio observations as well. Until recently, the necessary special efforts generally were not made to overcome these difficulties, resulting in undetected systematic errors in some cases. Mars, in contrast, is normally observed near opposition, and thus is viewed through the telescope in the middle of the night when conventional astronomical equipment can be used in more or less conventional ways. Also, Mercury, when available for visible observations, is only 5 to 8 arc-sec in size compared to Mars' maximum diameter of 25 arc-sec (see Fig. 6.3). Thus, even with equal efforts, our observational knowledge of Mercury would be very much less than that of Mars, a point also illustrated by Table 4.1. In actuality, the observation efforts have been overwhelmingly in favor of Mars.

Thus it is important to examine the "climate of opinion" concerning Mercury and inquire especially into how those attitudes came about. As in the case of Mars, such an inquiry can be valuable, not only as an aid in developing perspective for future planetary exploration ventures with spacecraft, but also as a means of gaining insight into the human process of discovery.

In the case of Mercury, we find no single dominant personality like Percival Lowell, but rather the incidental and often unrelated efforts of a number of investigators. Whereas with Mars there was too much emphasis on synthesis, hardly anyone has attempted that for Mercury. Whereas the need for consistancy and completeness in explaining the Mars observations drove Lowell toward unwarranted conclusions and rigidity, reports concerning Mercury were often characterized by blithe disregard of the obvious inconsistancy between the particular interpretation advanced and other published observations.

Another difference between the study of Mercury and that of Mars is that observations at radio frequencies as well as in the visible and infrared have played a dominant role compared to visual imagery. Thus the role of the planetary scientist is especially well illustrated as he deals with seemingly unrelated observations, at a variety of wavelengths,

## 6.4 Mass/radius diagram of Solar System objects

This diagram is a useful one to emphasize the discrete grouping of (1) lunar-sized bodies (E₁, J₁, J₂, J₃, J₄, N₁, S₆, and Mercury) and (2) asteroid-sized bodies, including all the other planetary satellites. The asteroidal-sized bodies class is gradational downward in size with increasing number, even to include meteorites and the minor debris of the Solar System and apparently constitutes a continuous fragmental population. Lunar-sized bodies, on the other hand, appear to constitute a discrete group, perhaps related to the circumstances of the formation of the Solar System.

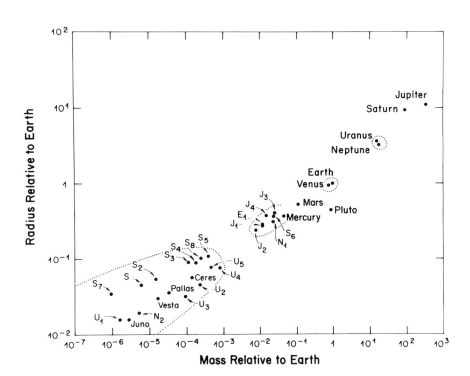

collected by a variety of observers using a variety of equipments with a variety of knowledge of associated observational errors. The modern explorer of planets, who would understand as well as merely record, must deal with remotely obtained measurements, most of which he did not collect himself. Thus he finds himself concerned primarily with distinguishing facts from fiction. When he encounters a supposed fact, he must use all his skill and ingenuity to determine its origin. Does the reported observation really mean what the observer thinks it does? Did the observer evaluate his own observational errors properly? Has the would-be explorer considered *all* the possible explanations for the supposed fact?

The last question is especially well illustrated in the case of Mercury because the one unifying characteristic of all the planet's investigators (including one of the present authors) was their failure to consider the possibility that Mercury might be rotating about the Sun other than synchronously, that is, with one face always toward the Sun as is

the Moon's rotation about the Earth. It was only the brute force discovery with radar by Pettengill and Dyce in 1965 of the nonsynchronous rotation that forced open previously closed minds. Even then, several years passed before the full implications of that nonsynchronous rotation were recognized for the supposed atmosphere of Mercury, and for the remarkable physical process involved in that rotation.

It is useful in that regard to refer to Fig. 6.4, in which all the tangible objects of the Solar System are plotted according to mass versus radius. There are some obvious groupings. Neptune and Uranus represent a similar pair, as do the Earth and Venus. Saturn and Jupiter are extreme in these parameters, as with many others. It can be seen that the gradational distribution of the asteroids overlaps at the upper end with many of the planetary satellites. An obvious distinct grouping, and one of great genetic interest, is that of the lunar-sized bodies, which seem clearly distinct from the asteroidal-sized objects on the small end and from the Earth-sized objects on the upper end. Mars in this plot would be regarded as transitional between the two groups. Mercury, however, is clearly a member of the lunar-sized objects, although far denser than the others, as we have already pointed out. Thus a first question which can be asked is— does Mercury look like the Moon? Traditionally it has been thought that Mercury does resemble the Moon. Drawings have contributed strongly to that impression. However, Mercury is so extraordinarily difficult to photograph or observe visually, as indicated in Fig. 6.2, that whatever pattern of albedo variations actually characterizes its surface, any rendition of those from Earth-based observations must necessarily be very degraded.

Thus, we may conclude that Mercury really could exhibit a most unexpected surface. For example, Mercury might display remnants of folded mountain belts which would be telltale signs of a molten, convective interior, or heavily cratered terrain left over from accretion, or perhaps even give evidence of having once undergone such intense solar irradiation that the outer, lighter layers of its crust were actually removed. On the other hand, Mercury might indeed resemble the Moon by exhibiting a surface comprised of dark circular maria with a few large craters emplaced in a lighter background of heavily cratered uplands. A little further reflection on the origin of the lunar maria, however, makes clear what an extraordinary result such a resemblence would be. It seems clear that

infrequent large asteroidal impacts created the mare basins early in the history of the Moon. These basins were then filled by a darker extrusive material chemically different from the upland material. Isotope dating of the mare rocks returned with the Apollo astronauts leaves little doubt that this filling process also ended early in the Moon's history, long before the present stratigraphic records on the Earth's surface began to accumulate.

Thus origin of the lunar "seas" required impact by especially large, and therefore infrequent, asteroids. The present distribution of these basins only on the Earth-side hemisphere may arise from the fact that the impacting bodies were part of a single event or family.

The filling process required chemical differentiation of some kind early in the Moon's history. That chemical "mix" must have been just the right kind to produce the mare/upland dichotomy. It is important to recognize that not just any molten silicate rock would produce this result. Mars, for example, does not appear to have undergone a similar process.

On the other hand, Mercury's density indicates that it must be of radically different chemical composition from the Moon or Mars and is perhaps closer to the Earth. And, it is almost three times closer to the Sun, where surely the asteroidal impact history would be very different. If some special asteroidal event were required to explain the formation of the mare basins on the Moon, what events could be hypothecated to account for mare-like features on Mercury! Thus, it would be most interesting if Mercury does exhibit a mare/upland appearance closely similar to that of the Moon. Indeed, we would be obliged to recognize that present concepts of the Moon's nature and history may be badly in error. More likely, the surface of Mercury recorded unique early events in the near environment of the Sun.

Other supposed similarities between the Moon and Mercury also warrant scrutiny. For example, even in the 1930s it was reported that Mercury's over-all albedo, color, and variation of brightness and plane polarization with phase all closely resemble that of the Moon. Yet the low albedo and slightly reddish-gray color are very likely the almost universal response of a great many kinds of powdered silicate materials to the intense solar irradiation experienced by an exposed surface. At Mercury this flux is an order of magnitude larger than for the Moon. Almost *any* dark powdered silicate surface will exhibit optical properties generally similar to that of the Moon and entirely consistent with the rather poorly determined optical data of Mercury. Thus, Earth-based optical observations (as well as the radar, radio, and infrared emission observations) really tell us a little more than that the surface of Mercury is composed mainly of finely divided silicate material. In another manifestation of the superficial evaluation of Mercury by generations of

scientists, during the last décade there was a vigorous, and in retrospect unwarranted, search for an atmosphere on Mercury. This search was motivated by what we now know are erroneous observations of the visible plane polarization across the tiny planetary disk and of supposed absorption of reflected solar radiation by a carbon dioxide atmosphere around the planet. There are, in fact, no valid observations indicative of any atmosphere on Mercury. However, very sensitive tests are possible from a flyby spacecraft.

So the supposed resemblence of Mercury to the Moon on the basis of telescopic observations should be regarded as superficial until close-up photography and other measurements from flyby spacecraft provide some really useful information about the surface physiography and morphology and any indications of a present or past atmosphere. Also, measurement of magnetic and electrical fields in the vicinity of the planet may provide an indication as to whether the overabundant iron required to explain Mercury's great density has accumulated in the center to form a large metallic core, like the Earth's, or is still uniformly distributed throughout the body. Thus just a single Mariner flyby of Mercury can provide the knowledge necessary to pluck Mercury from its mantle of ignorance and obscurity and place it within the genealogy of the Solar System. Is it a cousin of the Earth or a brother of the Moon? Or is it, perhaps, a unique object within the Solar System?

Particularly inexpensive opportunities to flyby Mercury occur a few times every decade or so when the Earth, Venus, and Mercury are all located in their orbits such that Venus provides a free gravity assist to a close flyby which propels the spacecraft on past Mercury. An opportunity of this type will be available in 1973 but not again until 1981; more costly launch vehicles are required at other times. The U.S. is planning a single launch to Mercury by way of Venus at the 1973 opportunity. As we will see again in the case of Jupiter, the limitation to exploring our companion worlds is not technological but is due to manpower and fiscal resources—and thus reflects national priorities. There will be only one "first look" at each planet and future histories will report this initial exploration of the solar system by spacecraft launched in the '70s. The 1973 Mercury/Venus mission will very likely be an endeavor in which the United States can take pride. As for cost, it's a little over 50 cents apiece for all Americans, payable over four years.

## 6.3 JUPITER—COMPANION OF THE SUN

One day our children's children, or their children's children, will send an automated spacecraft on a mission of discovery so fantastic that our present imagination can conceive of it only in terms of science fiction. This voyage will be to provide a "first look" at a nearby star—and especially at any planets which may be in orbit around it. One of the

great questions reserved for our descendants in the distant and uncertain future will be "how do we stack up?" Is our Solar System really nothing more than debris surrounding an average star among uncounted others? Or do we partake of uniqueness somehow as a system of planets, a uniqueness responsible for the development of intelligent life here.

Some may consider such thoughts to be nothing more than fragments of daydreams. Indeed, there is little that can be done now about the exploration of other stellar systems. And, we have no way of knowing when will come the fateful indication that "we are not alone." It may be in a year, a decade, a century —perhaps never. But the concept of exploring an unknown Solar System is nevertheless useful to us even now because the basic spacecraft technology and launch vehicles already exist to explore this Solar System from the innermost planet to the outermost. It is almost as easy to flyby Mercury as Mars or Venus. It costs no more to take pictures all the way to Neptune than to land robots on Mars. Thus we should be viewing this Solar System as we would an alien one—from the outside. We should recognize and try to overcome the foreshortened vision characteristic of the inhabitants of a minor planet of the central region who only now are beginning to gaze upward from their soil toward the cosmos.

If our Solar System were viewed from a distance, the most significant feature, aside from the Sun itself, would be the planet

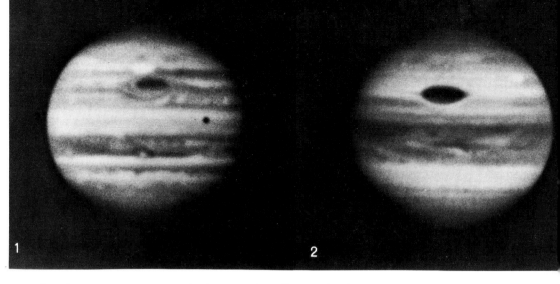

In these excellent New Mexico State University Observatory photographs taken in blue light the remarkable changes in atmospheric markings over several years are illustrated.

Jupiter. This giant among the planets contains more than twice the mass of all the other planets. Indeed, Jupiter truly can be considered the companion of the Sun because its mass falls just below the critical boundary between planets and stars; a bit larger and nuclear reactions would be initiated, causing it to be self-luminous. There is some evidence even so that it emits more energy as infrared thermal radiation than it receives as absorbed sunlight. As a source of radio noise, it is second only to the Sun, and the permanent magnetic field associated with the generation of that radio emission is the strongest yet encountered in the entire Solar System.

Even its satellite system is gigantic. The four largest, known as Galilean satellites after their discoverer, range in size from that of our Moon to larger than the planet Mercury. Furthermore, one or two seem to be composed of low-density material—perhaps even ice— whereas the others appear to be more rock-like in density. Yet all four exhibit pecularities in their reflection of sunlight which suggest surface conditions unlike those of the Moon or Mercury. Together, Jupiter and its satellites must be considered the most attractive targets for exploration in the whole Solar System, especially when some weight is given to our a priori ignorance of the planets, as illustrated previously in discussions of Mars and Mercury.

A less clear-cut question, however, is raised by the prospect of utilizing U.S. space resources to acquire pictures of Jupiter or its satellites.

The keenest concern for an early U.S. Jupiter probe has come from those interested in the origin of the intense radio emission and its connection with the interplanetary medium. Why divert resources into photography when there are so many other physical measurements that should be made in the near-planet environment?

The reason to photograph Jupiter from a space probe derives from the essence of exploration. Pictures of Jupiter at higher resolution than achievable from the Earth open the possibility for discovery of unexpected and unpredicted features. For example, the great red spot could not have been predicted from a priori knowledge of the planet; indeed its origin is still not well understood, but any theory of Jupiter's atmosphere must account for this remarkable phenomenon (see Fig. 6.5). It might have been difficult to argue persuasively that present ground-based resolution (about 1200 km for a low-contrast feature) would be sufficient to discover any significant features on Jupiter. Similarly, it is difficult to *prove* that a factor of, say, 3 to 10 or even 100 further increase in spatial resolution will be sufficient to lead to major new discoveries. Yet, experience so far in planetary exploration

103

from space probes suggests that a priori guesses concerning expected results should not be given excessive importance when a previously unobtainable step in spatial resolution or viewing is involved. The planners of the Mariner 4 experiment did not feel that the low resolution and extremely limited coverage of their experiment held much promise for discovery. Yet, it was sufficient to detect a lunar-like surface unexpected by most scientists. Luna 3's very-low-resolution view of the hidden side of the Moon could hardly have been justified in advance on its probable scientific return. Yet the absence of mare regions on the reverse face of the Moon is one of the space age's more important discoveries about the Moon.

In addition to the possibility of discovery of unexpected features, an imaging experiment of even only 3-10 increase in resolution and of high photometric quality should provide better information on the planetary limb darkening, on temporal variations of atmospheric features (if several pictures were acquired), and on cloud height variation as evidenced by an irregular terminator profile. The latter phenomena, of course, can never be viewed from Earth because we always see it from within 12° of the direction of the Sun. Higher resolution, and correspondingly larger amount, of photography hold even greater promise.

Similarly, low-resolution imaging of one or more of the Galilean satellites, even at only 10-30 resolution elements/diameter, should provide an indication of gross similarity or dissimilarity to the Moon as that object is viewed with the naked eye. Are there dark "spots" on a lighter background, as the Moon would appear if viewed at comparable low resolution, or are there really light bands, caps, and other nonlunar features, as some drawings based on visual observations have indicated? From even such crude information, a significant step in knowledge of these bodies could result.

Present U.S. space plans include a single launch toward Jupiter of a small spin-stabilized vehicle, Pioneer, in 1972 and again in 1973. Flight times would be about two years. How effective the exploration of Jupiter will be with this "low-cost" approach remains to be seen.

What does seem clear is that the U.S. possesses the basic technological capability to carry out a much more aggressive and diversified exploration of the planet and its satellites and to do so at a cost comparable to other planetary missions. The Grand Tour missions, presently being started, will provide an enormous return from Jupiter in the late 1970s. Then, we shall perhaps get to know this giant of the planets.

## 6.4 THE GREATEST VOYAGE EVER— THE GRAND TOUR

After a probe flies past Jupiter, where does it go? This is an intriguing question with a surprising answer. The probe can be made to go about as far out into space as might be desired, including right out of the Solar System! It all depends on how it passes by Jupiter— a remarkable consequence of the fact that both Jupiter and the probe are in motion about the Sun. During a near miss, energy is transferred from the giant planet to the tiny probe, energy enough to accelerate it onto trajectories simply beyond reach of any present or even foreseeable terrestrial launch vehicle. Indeed, that "first look" vehicle to go to a nearby star, hopefully to be launched by our foreseeable progeny, may well use such a technique to escape the gravitational attraction of our Sun, then steadily increase its speed toward that of light itself with the modest but continuing thrust of some not-yet-invented nuclear rocket.

In fact, Jupiter is believed to be interacting gravitationally in somewhat this way with natural interstellar probes—the parabolic comets. These comets, as the name implies, are observed to travel in nearly parabolic paths about the Sun, rather than the elliptical ones characteristic of all permanent members of this Solar System. Parabolic orbits can only mean that these comets have come from the most distant reaches of our own Solar System, or even the depths of interstellar space itself. Some of the parabolic comets happen to pass close to Jupiter and either lose energy to that planet to become captured in elliptical orbits about the Sun, or are accelerated onto hyperbolic trajectories to pass forever out of the gravitational field of our Sun.

In another natural parallel to our hypothetical interstellar probe of the future, some kinds of asteroids are probably ejected ocassionally from the Solar System after passing too close to Jupiter. In fact, giant Jupiter eventually dominates all who dare cross his orbit—either capturing the trespasser or expelling him from the Solar System.

Thus Jupiter is the keystone to truly extended space flight. Of particular interest for exploration is the possibility for a spacecraft to be so aimed that it will be accelerated onto a high-energy trajectory toward Saturn, Uranus, or Neptune, providing it arrives at Jupiter at just the right time. (This is the same approach that was alluded to in the discussion of Mercury; in that case, a close pass of Venus will, on some occasions, deflect a spacecraft in toward Mercury.)

In 1976, '77, and '78 it will even be possible to fly to Pluto. Less frequently, it is possible to swing by Jupiter in such a way as to pass *two* other planets. For example, Uranus and Neptune can be visited this way in 1978 and 1979, utilizing a second gravity assist at Uranus to acquire the needed energy to reach Neptune. Similarly, in 1976 and 1977 Saturn and Pluto can be visited.

Finally, every century and a half or so the major planets are lined up in such a way that it is possible to swing by Jupiter and continue to Saturn, then to Uranus, and finally on to Neptune. This fantastic voyage has been appropriately termed "The Grand Tour"

105

and can be executed on several occasions with launches in the years 1976, 1977, and 1978. The next set of opportunities for the Grand Tour occur around A.D. 2150! Thus there is a genuine argument of timeliness regarding this proposed space flight.

Another advantage of the swingby approach, besides the possibility of utilizing launch vehicles comparable to those scheduled for Mars exploration, is that the flight times are sharply reduced.

The swingby approach can reduce the flight time to Neptune from seventeen years for a direct flight to between nine and eleven years for a swingby one. The range of nine to eleven years arises because a choice must be made between going inside or outside the rings of Saturn. The shorter flight time can be achieved by passing inside the rings, close to the planet itself, but there is concern that the micrometeorite flux inside the rings may be dangerously high. Truly we are witnessing the end of science fiction and the dawn of scientific

106

### 6.6 Apparent sizes of the Outer Planets

In this composite prepared by the New Mexico State University Observatory, the outer planets Jupiter, Saturn, Uranus, and Neptune are all printed at the same scale. The effects of atmospheric turbulence can be seen to limit our knowledge from Earth of Uranus and Neptune severely compared to that of Saturn and especially Jupiter. Pluto would appear as an even smaller spot than Neptune since it is indistinguishable from a star. The Grand Tour missions can transform our enormous ignorance of these objects, and of their lunar-sized satellites, to a level perhaps comparable to our current knowledge of Mars.

### 6.7 Differing views of Saturn

The appearance of the planet, as recorded by New Mexico State University Observatory, is shown for changing orientation of the ring plane.

exploration of our Solar System when such a staggering prospect becomes a genuine engineering question, under study right now.

If the more conservative trajectory outside the rings of Saturn were chosen, a most favorable launch opportunity occurs in 1977. The probe would arrive at Jupiter 1.8 years later, permitting a vastly improved survey of that planet and of some of its satellites over that which may be obtained from the Pioneers F and G "first look" missions. There would be considerable analogy to the Mariner 4 (1965) and Mariners 6 and 7 (1969) missions to Mars in that the scientific data return probably will be orders of magnitude greater from the 1979 Jupiter encounter than from the Pioneer F or G mission.

Beyond Jupiter, it will all be totally unexplored territory (see Figs. 6.6, 6.7, and 6.8). But, fortunately, the data returned will still be considerably more abundant than in the "first looks" at Mars and Jupiter. Even at distant Neptune it is expected that at least 100 times the data of the Mariner 4 "first look" can be transmitted back to Earth.

The potential for discovery is hard to exaggerate. In each case, it would be possible to photograph the planet and its satellites, at greater resolution than is possible from the Earth, for long periods before encounter, providing a time record of atmospheric phenomena as the planet rotates under the camera's view. Other instruments aboard the spacecraft could measure also the infrared radiation emitted by the planet, especially from the nighttime hemisphere after the flyby, and gather information on atmospheric composition. Together such data can lead to major advances in the understanding of the nature of the atmosphere.

0350:49          0351:00          0351:10          0351:22

**6.8   An occultation of Saturn
by the moon**

This striking set of photographs was acquired by the New Mexico State University Observatory. Besides their aesthetic value, they provide a graphic demonstration of the difference in apparent size of these objects as seen from the Earth.

The satellites also are important targets. Even when they appear to the camera as only tiny points of light, their orbital motions can be studied, and their variation in reflectivity with phase angle can be measured. The latter can never even be attempted from the Earth because the phase angle never exceeds a few degrees—only 12 degrees maximum even for "nearby" Jupiter. The images of those satellites that pass close enough to be distinctly resolved should provide an especially significant result to compare with those of Jupiter's moons, our Moon, and Mercury. At present, it appears possible to flyby all six major satellites and photograph them at close range.

This same camera system which could play so vital a role in Man's "first look" at the outer portion of his Solar System may also be needed for a very important navigational task. At the enormous distances from the Earth involved in the Grand Tour, it is necessary to know quite accurately the direction from the spacecraft to the objective planet and its satellites, as well as the very precise range

and velocity information supplied by the radio tracking system. There is a direct connection between accuracy of angular knowledge and the amount of rocket fuel which must be carried to provide the trajectory corrections. Such precise angular information referenced to the fixed-star background could be provided by a properly designed imaging system.

Thus missions beyond Jupiter require for the first time that our robot have, and be able to operate, his own transit. In addition, the communication distances are so great that even radio waves take long intervals to travel between the spacecraft and Earth—up to four hours each way! Thus, there must be more on-board logic and automatic decision-making capability than previously. Indeed, the robot we send beyond Jupiter must be a more intelligent, more adaptive, and, hopefully, more farsighted machine than has been required to remotely explore the inner portions of the Solar System. Even so, it is likely that the cost will not be vastly greater though the total scientific return will be vastly larger.

352:05          0353:42          0354:15          0355:35

Thus we have already most of the necessary technology to not only explore the inner boundary of our planetary system guarded by dense Mercury, but also to leap beyond giant Jupiter to the very reaches of our stellar system. And this can be accomplished mostly within the decade of the 70s, utilizing only a small fraction of the U.S. space budget. Surely, no great nation could turn away from such a historic opportunity to expand the consciousness and vision of all mankind.

**REFERENCES**

Long, James E., "To the Outer Planets," *Astronautics and Aeronautics*, June, 1969.

National Academy of Sciences, "The Outer Solar System," 1969.

Newburn, R. L. Jr., "A Brief Survey of the Major Planets: Jupiter, Saturn, Uranus, and Neptune," Technical Memorandum 33–424, Jet Propulsion Laboratory, April 1, 1969.

"Planetary Navigation—The New Challenges," *Astronautics and Aeronautics*, May, 1970.

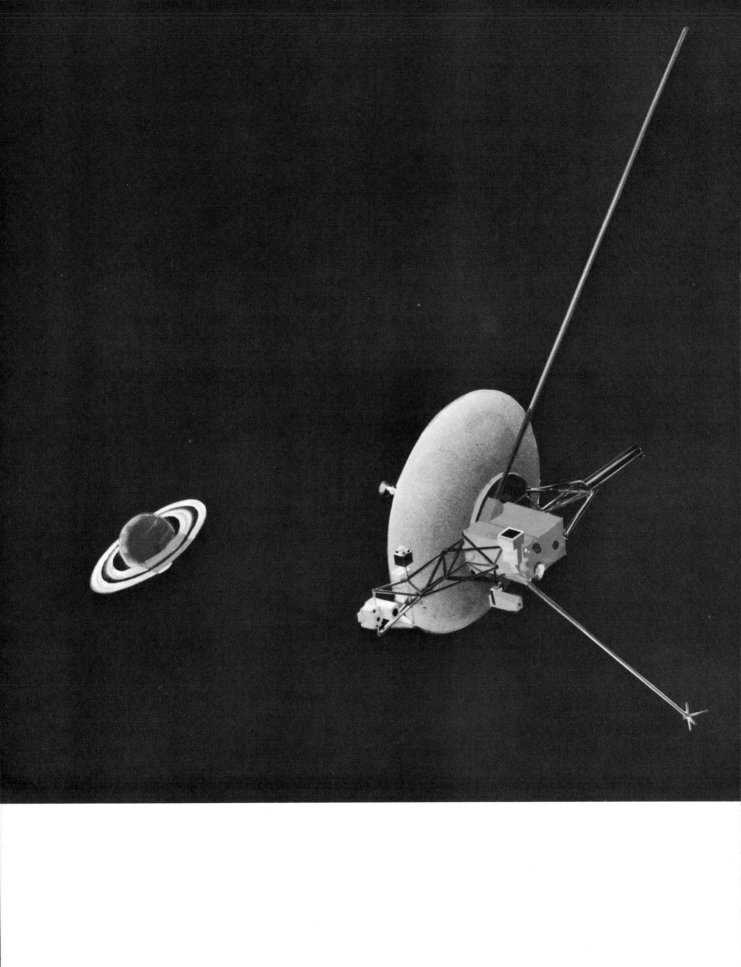

# CHAPTER 7
# The future

## 7.1 WHERE NEXT IN SPACE AND WHAT'S IN IT FOR THE U.S.?

In the late 1960s, an influential scientist called attention to the difficult decisions facing the United States in the post-Apollo period by asking two very good questions: "Where are we going in space, and what's in it for the U.S.?" Too often we lost sight of that second question, "What's in it for the U.S.?" In this context, "U.S." doesn't mean just scientists or industrial engineers, but rather all the citizens of the wealthiest and most powerful nation on Earth, citizens really beginning to wonder where they are going and why. Space exploration must make sense in these terms if it is going to survive as an attractive, progressive activity.

It is important to remember, of course, that the space program does indeed bring material benefits. Better world-wide communication systems are being developed; long-range weather forecasting is becoming possible; oceanic navigation has been improved. In brief, an enormously diversified advanced technology has become available. But these new services and capabilities, however valuable, represent only part of the results of space endeavors and should not be used as the primary justification. The real benefits come in the world of ideas—not just the scientists', but everybody's. It is there that the case should be made.

For example, the Ranger, Surveyor, Orbiter, and Apollo lunar photographs have probably produced an irreversible psychological effect on modern man. For one thing, the real Moon has been incorporated into his consciousness. Armstrong's first "small step" had to be photographed to be truly comprehensible to mankind. For another, Mars has been taken out of the realm of science fiction and brought within the scope of ordinary comprehension by the first tentative probings of Mariners 4, 6, and 7. In the coming decades this will be true for planets other than Mars. In space exploration we are witnessing a cultural change that affects not just Americans, but people throughout the world—a change perhaps as far-reaching as that of the last century when the theory of evolution forced a rearrangement of man's view of his place in time. Today man stands on craters first seen by Galileo. This profound consequence of the space program—the enlargement of human consciousness—should be accepted willingly and aggressively pursued.

Lunar and planetary exploration most directly manifest this cultural process so far. And exactly here we confront a profound challenge from the USSR, reflecting the genuine aspirations of the Soviet people. Profound because it is a challenge of ideas, universal in scope, international in impact. And now that the U.S. has made manned lunar exploration an American enterprise, the exploration of the planets (and Moon) by robots is clearly the next step in meeting the challenge.

111

The race to the Moon between the United States and the USSR should be regarded as more than just a muscle-flexing contest; indeed, it was a genuine cultural confrontation. Culturally significant results should be of even greater importance in the coming decade of planetary exploration. Style rather than scale, discoveries rather than exploitations, measure success in this confrontation. The United States has an unparalleled opportunity to shape the human mind and spirit in a crucial historic period. Indeed, the U.S. image as represented by the Apollo achievement is certainly a happier one than that resulting from our involvement in Vietnam, regardless of the final judgments of history about that tragic event. Exploration of the planets, especially as relayed back to the people of the Earth in photographs, is surely one of the more positive opportunities available to this nation.

In this section, we seek to establish criteria for decision making for the U.S. space program as a whole that will be more compatible with future uncertainties than the criteria used in the past. In the subsequent three sections, this rationale will be particularized for planetary exploration, lunar exploration, and Earth orbital activities. It is in that context that we shall discuss the future of space photography.

Any national decision involves a host of factors, among which scientific considerations are often (and correctly) of less rather than greater significance. The people making these decisions—the Congress, the President, the administrators of the agencies involved—have an almost impossible task of balancing apples and oranges. How does one assess the relative importance to the United States of Soviet exploits, U.S. technological development, scientific return, and unemployment in Denver? There must be guidelines if decision-makers are to do their job. There must be a strategy. What are the factors?

The *tangible benefits* from military and civilian space applications can be regarded in a straightforward manner. This is the least debatable part of the program. If such benefits are plausible, not necessarily demonstrated to be cost-effective, but merely plausible, the pragmatic spirit of the American people will usually support them. The problem arises when they are not very plausible or the return is very distant. *Domestic popular appeal* of space exploration is of paramount importance, and as this rises and falls there are going to be other effects.

The *commercial, industrial* and *technological* implications of space activities for the United States are very important. The decision to start Apollo meant a great dislocation of the U.S. engineering force in that whole new industries had to be invented and whole new classes of engineering developed. It was not a decision taken lightly. Similarly, the recent decision not to follow with another large-scale manned endeavor also means that some of that new capability will be lost to our over-all

**Table 7.1  Principal elements of U.S. space program in late 1960s**

| Element | Approximate effort, % |
|---|---|
| Manned lunar landing | 60 |
|    Saturn L/V development | |
|    Unmanned lunar exploration | |
|    Apollo S/C development | |
| Applications | 30 |
|    Civilian communication and weather | |
|    Systems | |
|    Military space program | |
| Earth-orbital astronomy | 4 |
|    OAO program | |
|    ATM portion of Apollo applications | |
| Planetary exploration | 3 |
|    Present and contemplated Mars efforts | |
|    Minor Venus effort | |
| Other | 3 |
|    Earth-orbital particle and fields | |
|    Miscellaneous | |

Summary of U.S. space strategy: 1961-69

1. Respond to Soviet challenge primarily with Apollo, developing a wide variety of space technology in the process. Apollo itself to be a single, technologically defined task.

2. Pursue military and civilian applications in accordance with plausible benefits and costs.

3. Pursue planetary exploration, Earth-orbital astronomy, and space science also, but without interfering with Apollo. Prepare for major expansions in the post-Apollo period to utilize the technology developed for Apollo. Response to Soviet competition in scientific areas not a major consideration.

technological activities. That also was not a decision taken lightly, but it had to be made as a result of irresistible pressure on the federal purse.

*The military and technological implications of the Soviet space program* are also of importance and, of course, also involve international prestige. Incidentally, the military space program is lumped together with civilian applications in this presentation, a more appropriate breakdown for strategy considerations than the usual military/civilian dichotomy. The *likely scientific value* of space activities is also important. It is not dominant, but it has some weight as part of the over-all scientific and technological activity of the U.S.

There is also *competition from other Federal programs for funds*. In recent years there has been much greater competition for funds within a set federal budget than previously when individual Congressional committees carried out rather independent reductions and additions.

Finally, *stylistic differences* between individual national leaders can be very significant. For example, the decision to go ahead with Apollo, the largest single decision of the U.S. national space program, was due in part to the way President Kennedy personally assessed the problem.

Given these factors, how has the United States committed its effort and resources? Table 7.1 gives a picture of the principal elements of the U.S. space program in the late

1960s. These numbers are only approximate but adequately demonstrate the general character and priorities.

The important thing to note is that high priority in space went to Apollo and to applications, especially military. Planetary exploration and astronomy received relatively minor emphasis. One can't give equal weight to all the different elements. Priorities had to be assigned, and that is what Table 7.1 illustrates.

Table 7.1 also gives some conclusions concerning the strategy implied by the distribution of effort. First, the Apollo Moon mission was considered a very good response to the expanding Soviet capabilities in space because it was a dramatic mission which the United

113

States had a good chance of accomplishing first. With anything less than Apollo there was a good chance the U.S. would lose. Second, the manned lunar landing presented such a complicated and broadly based task that it necessarily forced the development of many different kinds of space technology or permitted them to be developed in parallel with it. Most important, it offered a single, technologically defined task. It wasn't "do-good," it wasn't "look for life on Mars," but rather a specific job that this country could gear up to. This is very important. One of the main reasons the United States has had such difficulty in defining a satisfactory post-Apollo program is that it has not been able to find a technologically defined task of the same urgency and significance. The personal view of these authors is that Apollo will go down as one of the more important national decisions of the 1960s. The timing was right. It was the right kind of decision, the kind that made a lot of other things happen.

The second element of the strategy was to pursue space applications, both military and civilian, according to potential benefits and costs, estimated in some way or other.

The third element of the strategy, to be carried on without interfering with Apollo, was to do as well as possible in unmanned planetary exploration, Earth-orbital astronomy and space science, and some other minor things. But the main purpose in those areas was to prepare for a major expansion in the post-Apollo period that would exploit the technology being developed primarily for Apollo. In this case it was recognized that the United States could not challenge the Soviets before the completion of Apollo. As things turned out, the massive Soviet effort in planetary exploration during the 1960s failed to yield the expected results, and the smaller U.S. effort stole the show. On the other hand, the basic U.S. intention to exploit Apollo (and Saturn) technology in the 1970s for new Earth orbital and planetary ventures has also failed to materialize because of budgetary constraints. *Potential* Soviet lunar and planetary exploration ventures thus loom as a formidable challenge to the United States.

We can't have it both ways. Something had to be first and something else third. When ceilings are put on the top of a budget, cuts will have to be made at the bottom, and that is indeed what happened. It doesn't necessarily mean that NASA administrators, if they had their personal preferences, would have done it this way. This was the result of a national decisionmaking process. There were some pretty hard priorities, which, by and large, were correct. Apollo had to have first priority if it was to be at all. And the United States had to have Apollo to be in space significantly. It is hard to see how the United States could have gained its pre-eminence in space by any other path. In the 1970s we can make better use of our space resources if we can properly answer the question, "What's in it for the U.S.?"

## 7.2 SPACE PHOTOGRAPHY AND THE EXPLORATION OF THE PLANETS

This then is a broad view of what United States space strategy has been and where lunar and planetary exploration and space science has fitted in, the most important point being that there could not be an independent program for planetary investigation.

Surely one day man will go to the planets just as he has gone to the Moon, perhaps even for similar reasons. However, that time is still so distant that we will be concerned only with the unmanned exploration which will precede it.

In this regard, the Soviet program of unmanned planetary exploration called for an attempt to explore Mars and Venus at each launch opportunity starting in 1960 using the most advanced booster available (see Table 4.2). Apparently their planetary program was deemed truly important as it experienced little interference from the rest of their program. Their many failures would be interpreted as a rather poor job by U.S. standards. Nevertheless, their perseverance was rewarded finally with success in sampling directly the atmosphere of Venus; Venera 4, 5, 6, and 7 have re-established the USSR as a contender in planetary exploration.

It is useful here to assess the factors that affect planetary exploration decisions. The first consideration is that of *scientific achievements*. The primary rule of space exploration is to design and build payloads so as to maximize the potential scientific return from the mission. This rule has been recognized in the United States and is playing a greater role in each successive planetary mission as engineering confidence has increased. This is dramatically illustrated by observing the data return from the successive Mars missions Mariner 4, Mariners 6 and 7, and Mariner '71 (see Fig. 1.1). The cultural values that accrue from planetary exploration come specifically from scientific results and, in a more general way, from bringing into man's consciousness the nature of other worlds in the solar system. Hence, *scientific achievements* indeed are important, much more important in planetary exploration than, say, in Apollo.

A second factor that the United States must examine is *actual* and *potential Soviet endeavors*. For example, in October 1967, U.S. engineers were studying a capsule mission to Venus in 1972. Those efforts were abruptly terminated when the Soviets carried out such a mission. It has been repeated four times since then by the USSR. There were things wrong and some loose ends, to be sure. But suddenly the U.S. program was changed. Today, the United States cannot make the same kind of unilateral decisions for proposed Venus missions because of the long leadtime of all space missions; the Soviets will have had two or three more opportunities to make new investigations of the planet. Moreover, it's likely a similar situation will emerge

115

regarding a Mars lander, especially as the U.S. has postponed its first attempt from 1969 to, at earliest, 1975. However, the Soviets may yet have difficulties because they have never maintained communication with a spacecraft at Martian distances, suggesting a real weakness in their basic design.

Concerning U.S. planetary-exploration decisions, we need to consider the question of *sterilization policy* and *practice*. The United States has insisted upon, and maintained faithfully within its own program, an extremely rigorous attitude toward sterilization. The Soviets have not. Unilateral sterilization policies in space make as little sense as do unilateral conservation policies on Earth.

A third factor that affects planetary exploration decisions is *launch vehicle availability*. This factor has dominated planning in a peculiar way. Initially U.S. missions were limited because only the Atlas/Agena was available. But just as soon as a suitable launch vehicle, such as the Atlas/Centaur and Titan III-C, became available, there was a strong effort to assign important planetary missions to Saturn-class boosters. When there were not enough boosters, planetary exploration stood hat in hand. When larger boosters became available, unmanned planetary probes were to be launched by huge costly rockets whether appropriate or not. In 1965 the approved planetary program called for stepping directly from the Atlas/Agena capability to that of the Saturn V (a factor increase of 100

in payload and perhaps 25 in cost), a move which reflected extreme launch-vehicle considerations to the exclusion of economy, efficiency, and good science.

*Uncertain future funding* is an enormously important factor in planetary exploration. The technical lead times are very long and planetary launch windows simply cannot be slipped to meet technical delays. The Soviets operate on the basis of a five-year program. They lay out certain objectives, and presumably renegotiate their budgets annually, but their basic expenditures are pretty well known for a number of years in advance. By contrast, we have an annual cycle. The U.S. planetary program can be completely renegotiated every year—an inefficient and expensive way to operate because of so many false starts. Moreover, it weakens momentum and morale.

Finally, among Congress and the Administration and the Space Agency there necessarily develops a somewhat protagonistic relationship which can lead to the freezing of positions. The Agency cannot afford to rewrite the script every year, even though events change. So there is an agency "time-constant." National funding also has a time-constant. These two can get out of phase. The result is a program of limited flexibility. If we are going to compete effectively in planetary exploration, yet within limited funds, then we need to be as flexible and aggressive as we can. This situation is doubly ironic in that our annual budget cycle should, in principle, provide greater flexibility to pursue an opportunistic program, thus counterbalancing some of the advantage of the Soviets with their sustained support.

These then are some of the factors, the more important ones, which affect planetary-exploration decisions.

Another consideration is that of alternative scientific approaches. One approach, originally formalized in October 1964 by the NAS Space Science Board, the search for life on Mars, is to be the single unifying element of planetary exploration. Although the early U.S. planetary program did develop flights to both Venus and Mars (in 1962 and 1964), there has been only a single launch toward Venus since then (in 1967), and a single additional flyby (on the way to Mercury) is now planned for early 1974. No others are presently approved. In contrast, two new flights went to Mars in 1969, and two more are scheduled for 1971; yet another two for 1973 had been approved previously but were postponed until 1975 due to financial pressures.

NASA clearly has created a Mars-priority planetary program generally consistent with the 1964 SSB policy. Other planetary missions are also solicited, but only as add-ons to the basic Mars program. Thus, when resources are limited, the U.S. planetary program has necessarily become practically a "Mars-only" program.

Another scientific approach emphasized the importance of first looks at new worlds. This "First Look" approach proposes going to each more distant planet as soon as possible and then following up with more specific missions based on what is found in the first look. For example, the program to swing by Venus onto Mercury in 1973–74 with a single Mariner spacecraft is consistent with this objective— but why not two launch attempts for this different and exciting exploration venture? There is also a program to launch a small Pioneer-type spacecraft toward Jupiter in 1972 (for arrival in 1974) to be followed by another a year later. It is hoped that much more substantial programs will soon get underway, especially the Grand Tour and related flights which must be launched in the 1977–79 time period.

The "First Look" approach means going from the science-fiction stage of thinking to the frontier of reality for each planet as quickly as possible . We can worry about the question of life on Mars when we know a little more about Mars, particularly after we have found out whether or not it is even an interesting question. Maybe Jupiter is a better place to look, or maybe there are other ways to study the problem of life besides looking for it on the surfaces of the planets. This then, is an adaptive approach to exploration.

The "First Look" approach concentrates, moreover, on exploratory measurements that can give unexpected information. Photography is emphasized as being the "widest bandwidth" and most unbiased experiment possible in many cases. It raises fruitful questions, not

117

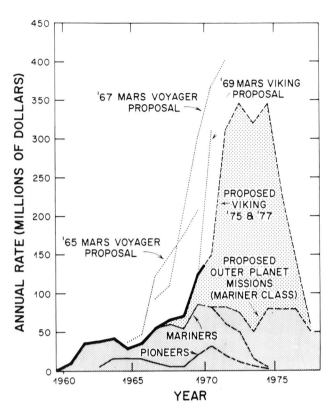

The actual funds appropriated from 1960 to 1970 for planetary exploration are shown by solid lines. Dashed lines from 1970 onward show the projected expenditures for those programs approved by about the middle of 1970 and also an early estimate of the cost for a Mariner-class Grand Tour group of missions. It can be seen that the Viking Lander/Orbiter program dominates the projected funding. Earlier unsuccessful attempts to initiate major Mars lander programs are shown with dotted lines and include: (1) the 1965 proposal to use Saturn 1B/Centaurs for missions in 1971; (2) the 1967 proposal to use a Saturn 5 for a mission in 1971; and (3) the 1969 plan to use Titan 3C/Centaur for 1973 missions. The present Viking plan is the same as (3), except the date has been postponed to 1975. This diagram illustrates well the effect of sophisticated lander programs on planetary budgets. However, the projections beyond 1970 will undoubtedly be out-of-date by the time of publication due to continued readjustment in NASA plans.

just answering them. Photos can suddenly open up the meteorology, geology, and history of the planets. But there are other incisive "First Look" exploratory measurements too. For example, just a temperature measurement deep in the atmosphere of Venus proved to be a very good exploratory tool, because it distinguished among a host of possible Venuses. Suddenly the spectrum of hypotheses narrows to a "real world" picture. That's what exploration is: the narrowing down of alternatives, not the answering of a specific question the way one does in a laboratory.

Intertwined with this question of scientific strategy has been the question of the level of support for the U.S. planetary program. In general, unmanned missions to land scientific payloads on the surface of the Moon or planets generally seem to cost a factor of at least three more than photographic flybys or orbiters. Hence the "Search for Life on Mars" strategy necessarily implies a considerably higher level of support—once survivable landers become involved—than is required for the "First Look" strategy. Figure 7.1 compares proposed Mars lander efforts with actual funds awarded for all planetary exploration over the last decade. However, a number of more ambitious and more costly Mars lander proposals have not materialized. The authors of this book feel that such expansions were neither necessary nor desirable from the point of view of the question, "What's in it for the country?" There remains, however, the fact that fiscal strategy has been concerned primarily with promoting a sharp increase in the level of funding in planetary exploration rather than maximizing the over-all exploratory return within a modestly expanding planetary budget. Were this not the case, there is little doubt that a simple, pioneering Mars atmospheric probe/lander could have been attempted by the United States in 1969 or 1971. Instead, the first U.S. landing effort, albeit a more ambitious one, will not take place before 1976.

Thus, the planetary program remains the focus of divergent fiscal and scientific strategies. Much of this controversy and rhetoric would abate if the participants were required to straightforwardly answer the question, "What's in it for the country?" One must hope that the current emphasis on limiting federal expenditures will lead to a healthy and trim planetary program rather than to an unbalanced and inefficient one.

## 7.3 PHOTOGRAPHY AND THE FURTHER EXPLORATION OF THE MOON

During the last several years lunar exploration has undergone a profound change. Very large amounts of color and black and white film photography, as well as live television, have been returned both from orbit about the Moon and from the lunar surface itself. Most important, considerable amounts of lunar soil and surface rock fragments have been minutely analyzed for their isotopic, elemental, and mineral abundances, and for their textures at all scales. Indeed, the early Apollo samples have been the most thoroughly examined substances in the history of mankind! Thus, in the course of less than a decade, scientific knowledge of the Moon has leaped from a level comparable to our current knowledge of Mars to a degree of familiarity approaching

that of isolated terrestrial localities such as central Antarctica. Mankind today possesses more information about and understanding of the lunar "seas" than of the isolated depths of some terrestrial oceans.

As a result of this unprecedented human endeavor, the character and pace of lunar exploration in the next decade must surely change from that of the 1960s. The very success of the Apollo program in demonstrating the superiority of the American space efforts, and by implication the superiority of the U.S. industrial system, has reduced the U.S. national priority for subsequent efforts. The effect on Soviet space priorities remains to be seen. At present, the United States is in a "spendout" phase of lunar exploration, flying the left-over Apollo hardware at about as slow a pace as is practical and safe. This slowed-down procedure constitutes a boon for the scientific study of the Moon in that the reduced pace permits the results of one flight to be incorporated more effectively into the location and mission profile of the succeeding flights—that is, the exploration process becomes more truly an adaptive one.

Lunar exploration beyond the present "spendout" phase remains uncertain. Since the cost of a single Apollo mission is comparable to half the annual budget of the National Science Foundation or to a quarter of that of the National Institutes of Health, it is difficult to justify continued manned lunar exploration principally on scientific grounds. Such sums

119

are large even if considered justifiable in terms of national prestige. Thus U.S. manned lunar exploration may continue to decline in pace and perhaps even reach a hiatus of some years duration. Renewed national rivalry in space might then again stimulate U.S. manned lunar efforts. On the other hand, without such nationalistic stimuli, lunar exploration might conceivably lead eventually to genuine international cooperation in space. In the case of Antarctica, the nationalistic "race for the pole" was followed by a lapse of nearly half a century and two world wars until adequate transportation technology had become common and the International Geophysical Year had provided a worldwide scientific framework for cooperative efforts. On the other hand, should a "race for Mars" develop the position of the Moon as a training ground, a permanent base, and perhaps as a forward staging or refueling station for the Mars mission could assure it a permanent place for human abode.

Whatever the pace of lunar exploration, photography will continue to play a major role. Man will continue to want to "see" the lunar terrain as well as the astronauts carrying out their tasks through live television and returned film. The collection of samples and other kinds of surface study by astronauts must be carefully documented by conventional and stereo photography. And extrapolation of the knowledge from those isolated localities to the remainder of the lunar surface will be carried out principally by specially designed photographic techniques from the orbiting command ship supplemented by remote sensing of wavelengths outside the visibile.

The Moon will continue to become more familiar to Man through photography, continue to become part of the consciousness of a larger and larger fraction of mankind, until ultimately it will seem like just another isolated part of the Earth. In the meantime, "What's in it for the country?" (and, "What's in it for the Soviet Union?", as well) will determine how fast—or how slowly—the exploration of the Moon will be completed.

### 7.4 PHOTOGRAPHY AND EXPLOITATION OF EARTH ORBIT

"What's in it for the United States?" has been the formal guide to NASA's programs for Earth orbital applications throughout the 1960s. As a result, communication satellites and weather satellites are now taken for granted by the American people. Significantly, space photography is a major aspect of one of these two principal areas of space applications and, therefore, will continue as a major endeavor in Earth orbit throughout the next decade.

The 1970s, however, will see two new dimensions added to Earth orbital space activity. First, systematic observation of the Earth's surface itself from orbit will become a practical "cost-effective" endeavor in which photography will be the principal tool. Second,

Earth orbit will become the arena of the activities of many nations, perhaps even of the U.N. itself, as a result of the proliferation of launch-vehicle technology throughout the industrialized nations of the world. Thus the decade of the 1970s should see Man's awareness of his own planet—and of his modifications of it—expand enormously through the dissemination and diversity of Earth orbital space activity, especially that involving photography.

Photography from orbit can be valuable even at rather low resolution. For example, meteorologists commonly use low resolution views of cloud cover acquired frequently over large areas for synoptic studies rather than higher resolution pictures of smaller regions acquired less frequently. Indeed, a principal limitation to the usefulness of weather satellite data is on the ground, not in space. Somehow the enormous mass of imagery must be processed *and analyzed* in a timely manner to be useful in weather forecasting. Hence, ground resolution must be traded off against geographical and temporal coverage. There is considerable potential importance in semi-automatic methods of processing and analysis to accelerate the flow of data beyond the limits presently set by the capacity of humans to view and interpret cloud photography.

Similar considerations apply to future satellite systems which will monitor the condition of the oceans, the distribution of polar ice, and large-scale air and water pollution throughout the world. A valuable by-product and occasional extension of such studies will be abundant low resolution coverage of the fixed features of the Earth's surface, permitting improvements in global cartography, physical geography, and regional geology.

High-resolution photography from Earth orbit would permit the observation directly of the works of man and nature—the status of crops, forests and grasslands, the nature and growth of urban areas and distribution systems, and the development of local areas of pollution.

Although all of these man-related uses of high resolution Earth orbital photography generally require repeated coverage of the same areas of the Earth's surface in order to recognize changes, the frequency of such repetition will usually be much less than that of weather satellites. In any case, the areas to be covered generally will be but a small fraction of the Earth's surface. Thus the total data flow can be kept within the limits of ground data processing and analysis. As in the case of global low-resolution observations, information on the fixed features of the Earth's surface will also be obtained making possible selected mapping of local terrain and associated geological features.

However, the potential benefit accruing from such diverse and multinational exploitation of earth orbit in the coming decade carries with it the same dangers and challenges characteristic of "new frontiers" on the Earth's surface. The exploitation of space requires

121

acceptance of the concept of open skies even when rivalry for potentially valuable resources might be involved. Thus the future may bring complex interdependencies and rivalries in space.

More profound may be the challenges to the governing and administrative institutions on Earth. How can information concerning crop conditions in, say, Illinois, made accessible by a U.S. satellite monitoring system, be relayed effectively to those affected? Will they know what to do with the information? Who should pay for the service? To extend this hypothetical problem further, suppose the observed crops are in Afghanistan? What are the considerations involved for the American government? What's in it for the U.S.?

Thus space is intrinsically international. Multinational exploitation will not be easily managed through the traditional institutions of local and national governments. The challenge of exploitation of earth orbit will represent in pure form an opportunity for the peoples of the world to work together for their common good, to keep their planet habitable and bountiful.

The rate of dissemination during the coming decade of "the view from space" to mankind generally as a result of earth orbital photography will be both an index of and a catalyst for the growth of a sense of unity among mankind. Perhaps if Homo sapiens can recognize and accept the challenge of exploitation of earth orbit, he will thereby demonstrate his capacity to rise above his past, to tame his runaway technology, and to make his Earth a noble place from which to reach out ultimately into the universe itself in search of worthy companions.

# APPENDIX

# A. Photography as a communication process

A photograph is an intermediate recording and filtering of the brightness distribution in the scene viewed by the camera. When that representation of the visual scene is, in turn, viewed and interpreted by the combination of the human eye and brain, then and only then does a photograph have meaning. A photograph thus can be regarded as nothing more nor less than a link in a communication system. This is a particularly appropriate basis on which to consider planetary photography from space probes because the intrinsic difficulty and expense of returning photographic information from planetary distances demands a most careful review of what signals really need to be transmitted by the communication system. What is it we wish to know about the visual scene as viewed from space? How can that information be most effectively recorded and returned to the ground? How can the view then be reconstructed to relay to the human observer all the information actually transmitted? These are the basic questions that have confronted those who have photographed the Moon and Mars from space.

## A.1 WHAT IS A PHOTOGRAPH?

Not only is a photograph a representation of the scene viewed by the camera, but it is also a record on some sensor of the actual light intensity distribution in the focal plane of the camera. The imaging process itself involves some distortion of the incident radiation; uniform plane waves collected by the optical system will be imaged on the sensor as some small, but finite, bundle of light rays, a process described by the notion of a point spread function. Thus, the photographic record of the scene intensity distribution is one in which the fine detail has been smeared out somewhat by the point spread function of the optical system itself. However, the recording by the sensor of that image intensity distribution contributes further blurring. Thus, the eventual "output" of the sensor, be it voltage, transmissivity, or other physical quantity, will exhibit the combined blurring of both the optical system and the sensor. Such angular effects are important to the question of resolution which will be discussed later.

It is also important to consider the sensor as a light intensity transducer and to consider its signal and noise characteristics. Any small area of its surface can be considered to have some responsivity, $\rho$, which relates the output physical quantity, $V$, to the incident light intensity in the image plane, $i$. Hence, the signal output will be $V_S = \rho \cdot i \cdot a$, where $a$ is the reference small area of the sensor involved.

The equivalent noise input, $\delta i$, on the other hand, may have some (perhaps complex) dependence on both the absolute area of the sensor involved and on the incident intensity. Hence, we can state, functionally at least,

$$\delta i = f(i, a) \qquad (1)$$

and

$$V_N = \rho \cdot \delta i \cdot a \qquad (2)$$

where $V_N$ is in units of equivalent signal output. Since $\rho$ presumably is known from calibration data, we can alternatively speak of $i$ and $\delta i$, the reconstructed image intensity and its associated noise. The output signal-to-noise is simply $V_S/V_N$. In space photography there are generally a number of extra links in the communication chain associated with returning the data to Earth, and noise may creep in along the way. Therefore, we include in $\delta i$ all forms of noise arising between the original image plane and the finally reconstructed view on the Earth and reference them in terms of equivalent input intensity.

Thus a generalized photographic system may be regarded as a remote viewing system for an Earth-based observer, generally with time delay, in which: (1) the scene viewed undergoes angular blurring arising from the camera optics and the sensor; (2) the light intensity corresponding to each small element in the viewed scene is transduced into some measurable physical quantity and is ultimately displayed as a related shade of gray, and; (3) spurious variations in gray shades, that is, noise, are also present.

## A.2 BITS AND PIECES

If any photograph is examined at progressively increasing magnification, the variations of gray level within the area under observation will be found to become indistinguishable as the area is made smaller and smaller. We denote tentatively an individual small area within which only a single intensity can be discerned as a pixel (shorthand for "picture element"); a photograph can be regarded as composed of a large number of adjacent pixels. Later we will refine the notion of a pixel. If each pixel is taken to be $1/K$ mm in size, and the photograph is $A \times B$ mm in dimension, then the number of pixels in the picture, $n_P$, is

$$n_P = K^2 AB \qquad (3)$$

124  However, the intensity corresponding to an individual pixel $(m,n)$ can be determined only with some limited accuracy due to the noise from the original sensor, or from subsequent steps in the communication process. At this point we wish to tie our analysis directly to the notions of communications by introducing the concept of a bit, the unit of binary encoding. Bits can be used as units of measure of the combined number of pixels and gray shades in a photograph. They provide a most suitable reference as to the communications requirement of a photograph.

Specifically, we wish to know how many zero-or-one characters in a binary word would be required to encode the sensor output $V_{mn}$ for transmission, that is, how many distinct levels will be available to encode the output of each pixel. There is, of course, no exact way to answer such a question, since the more precisely $V_{mn}$ is encoded, the more accurately, in principle, the original intensity $i_{mn}$ can be inferred in the presence of the uncertainty $\delta i_{mn}$. But we note that, in reality, precision of encoding beyond a certain point helps mainly to provide improved knowledge of $\delta i_{mn}$, and only indirectly of $i_{mn}$. Thus, efficient encoding practice often is based on the notion that the uncertainty in $i_{mn}$ resulting from the limitations of encoding should be roughly comparable to the intrinsic uncertainty in $i_{mn}$ itself, $\delta i_{mn}$.

Hence we wish to relate the signal-to-noise per pixel to the number of levels per binary word.

In this case $n_b$, the number of bits in each binary word, is related to the signal-to-noise by

$$2^{n_b} = 1 + \frac{i}{\delta i} \qquad (4)$$

or, the number of bits/pixel, $n_{mn}$, is

$$n_{mn} = 3 \log_{10} \left[ 1 + \frac{i}{\delta i} \right] \qquad (5)$$

This must be viewed as a very simplified approach to estimation of bit requirements of a photograph. Yet it does permit us to estimate the number of bits in a photograph with about the same accuracy as one can refer to the number of gray levels present. Thus the total number of bits in a given photograph, $n_T$, can be stated approximately as

$$n_T = n_p n_{mn} = 3K^2 AB \log_{10} \left[ 1 + \frac{i}{\delta i} \right] \qquad (6)$$

## A.3 WHY WANDER INTO THE SPACE FREQUENCIES DOMAIN?

In many observational systems it has been found to be useful to analyze the temporal frequency content of the output signal. An analogous situation exists for photographic systems concerning the frequency distribution of the *spatial* variations in intensity which constitute the original and reconstructed images. There is, of course, only very rarely any physical significance to a particular space frequency in a scene. However, white light photographic systems (temporally incoherent) are reasonably linear in intensity in the sense that the intensity distribution in the image plane is the arithmetic sum of the intensity distribution arising from each small element of the quasi-random (spatially incoherent) intensity distribution of the scene being viewed. Under such conditions, the image degradation of any scene, in principle, can be accurately predicted from knowledge of the point-spread function of the optical system.

However, the actual usefulness of the point-spread function is limited due to practical computational difficulties; instead the Fourier transformed functions, which are separable, have proven to be more important. For example, the line-spread function for each of a number of optical elements can be considered together as $l(x')$.* Denoting the original intensity distribution in the scene being viewed as $o(x)$ and that at the corresponding point in the image plane as $i(x')$, then**

$$i(x') = \int_{-\infty}^{\infty} l(x' - x) o(x) \, dx \qquad (7)$$

By comparison, if $\omega$ denotes spatial frequency, the cumbersome term $l(x')$ can be represented instead as a space-frequency filter function $T(\omega)$,

$$T(\omega) = \int_{-\infty}^{\infty} l(x' - x) \cos \omega x \, dx \qquad (8)$$

Then

$$l(\omega) = T(\omega) O(\omega) \qquad (9)$$

and

$$i(x') = \int_{-\infty}^{\infty} T(\omega) O(\omega) \cos \omega x \, d\omega \qquad (10)$$

where the capital letters signify the Fourier cosine transforms of the lower-case intensity functions. The usefulness of the space-frequency approach is obvious when it is recognized that if sinusoidal (or even square wave) bar charts are used to investigate optical performance, then the observed contrast reduction of the image compared to the target is a direct measure of the attenuation of the space frequency in question:

---

* Assuming a symmetrical line spread function.
** Somewhat simplified from the one-dimensional derivation of Smith (1963, p. 339).

## A.1 Processing of Mariner 4 television picture

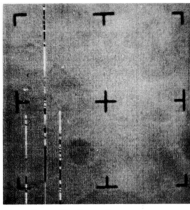

Far left: raw data of frame ten from Mariner 4 as received in July 1965. Next photo: same data, after several days of processing with preliminary enhancement and with torn lines corrected and fiducial marks and spurious interference removed by interpolation. Far right: finally calibrated and enhanced photograph with most electrical noise removed. To its left: same, except processed with high-space-frequency filter. The last two steps were not completed until more than a year after receipt of data.

$$T(\omega) = \frac{M'(\omega)}{M(\omega)} \qquad (11)$$

where $M$ is the modulation and is defined as $(i_{max} - i_{min})/(i_{max} + i_{min})$. The prime refers to the image plane observation. $T(\omega)$ is known as the modulation-transfer function ($MTF$) and can be used to describe the angular blurring properties of the sensor. However, sensors normally are not completely linear over the entire intensity ranges involved, and it must be presumed that the output signal $V_{mn}$ has already been converted back to the apparent image intensity $i_{mn}$ by means of the known calibration $\rho(m,n)$ before the contrast reductions are considered.* Thus, if we denote the $MTF$ of the optics as $T_o(\omega)$, then

$$l(\omega) = T_o(\omega)T_s(\omega)O(\omega) \qquad (12)$$

Thus the recourse to the space-frequency domain is a useful way to measure the angular blurring effects of any photographic system. Equally important, it offers a way to correct for at least some of those effects. Equation (12) is separable. Therefore,

$$O(\omega) = \frac{l(\omega)}{T_o(\omega)T_s(\omega)} \qquad (13)$$

and**

$$o(x) = \int_{-\infty}^{\infty} O(\omega) \cos \omega x d\omega \qquad (14)$$

## A.2 Processing of Mariner 6 television picture

Below: raw analog data of near encounter frame 18 from Mariner 6 as received in August 1969, including on-board enhancement effects. To its right: electrical noise to be removed from raw data. Next photo: raw data with noise removed and then enhanced, but without correction for vidicon response. Far right: same, except processed with high-frequency filter. All these processing steps were carried out within a few days of receipt of data.

* See "effective image" concept discussed by Brock (1965, p. 45).
** We continue to use only the one-dimensional cosine transform for illustration. Complete solutions generally require two-dimensional formulation of both sine and cosine transforms, although this complexity is not required for all applications.

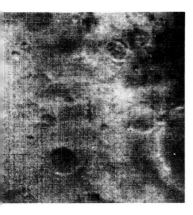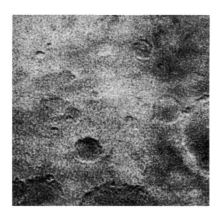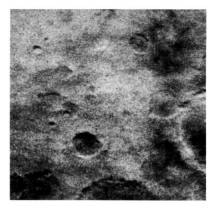

In principle at least, the image-intensity distribution $l(x')$ can be transformed to the space-frequency domain, the resulting coefficients $l(\omega)$ increased by $[T_o(\omega)T_s(\omega)]^{-1}$ to yield estimates of $O(\omega)$, and that result transformed back to yield an estimate of $o(x)$. The limitation to such a process ultimately will be the sensor or other image noise, $\delta i(x)$, which will also be increased by $[T_o(\omega)T_s(\omega)]^{-1}$ and ultimately will make the reconstruction process unmanageable. Successful application of such filtering to improve resolution has been carried out extensively with lunar and Martian television photographs where (1) the signal-to-noise for each pixel was generally high, (2) the data were already in a form well suited for processing by digital computers, and (3) electrical interference also needed to be removed. More extensive use of filtering in the reconstruction

of television pictures can be expected as the techniques for doing so become better known.

Up to this point, our interest in filtering has been merely to help reproduce the original scene as faithfully as possible. However, the eye-brain combination can sometimes extract more information from the photograph if the contrast change in the reconstructed image has been artificially enhanced, that is, by use of high gamma processing procedure with film. Similarly, it is sometimes desirable to artificially "sharpen" the detail in the picture at the expense of the low space frequency information. Both kinds of extreme enhancement have been carried out with the Mariner 4, 6, and 7 television pictures of Mars in order to convert the precise intensity data into information about the extremely low contrast scenes, as illustrated in Figures A.1 and A.2.

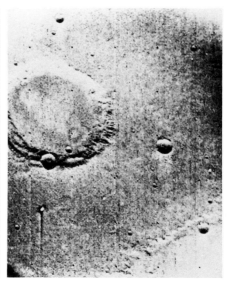

These examples again lead us to the basic connection between angular resolution and the signal-to-noise of individual resolution elements. This, indeed, is one of two unifying concepts underlying the whole complicated process of design and utilization of photographic systems in space. With it, the system parameters can be adjusted favorably to maximize the useful pictorial data transmitted back from space.

## A.4  RESOLUTION AND CONTRAST

After the space photographer has made sure he is communicating back from space all the bits he can justify, he then must wrestle with the question of how to expend them most effectively. Should he go for a few isolated very-high-resolution pictures, or lots of contiguous low-resolution coverage, or, as usually is best, some mixture of these extremes? Thus a clear understanding of the concept of resolution is necessary. In the testing of optical systems, resolution is a measure of the ability of the system to render barely distinguishable, to a human observer, a standard pattern of black and white lines.

However, the resolution defined in this way obviously depends on the contrast of the bar chart. If the spaces between the lines are only twice as bright as are the bars, that is, a contrast of 2:1, then the barely detectable line spacing will be coarser than if the contrast were 100:1. Hence the limiting resolution of photographic systems often is stated as a certain number of lines per mm at a specified contrast. Contrast is used here specifically to mean the ratio of the maximum intensity to the minimum across a bar chart, or across a sinusoidal variation pattern.

Returning to our discussion of a pixel, there must be a connection between the area of a pixel, with its associated intensity uncertainty, and the limiting resolution vs. contrast relationship. Since the output signal-to-noise from the sensor depends on the actual area involved, certainly a high contrast image pattern can be recognized under conditions of worse signal-to-noise than can a low contrast pattern. Therefore, we should expect on very general grounds a relationship between resolution and contrast qualitatively similar to that observed. Conversely, we can sharpen up our concept of a pixel by identifying its area with a resolution element corresponding to a particular contrast, and then choosing a value of $S/N$ appropriate for the same contrast. Yet again we find ourselves dealing with intrinsically imprecise concepts, because resolution ultimately depends on the threshold of eye-brain recognition of a standard pattern. It is not possible formally to relate the physically meaningful term "signal-to-noise" to what is basically a psychophysical process. Thus, again we will have to fall back upon engineering experience and use what empirical information may be available.

For this purpose we define $R$ in line pairs per mm as the limiting resolution of an imaging system determined by the visual viewing of a series of three-bar targets to determine the most closely spaced detectable pattern. $R$ is obviously a function of the contrast of the pattern. We then imagine that the same image used to determine $R$ is scanned with a (noiseless) microphotometer whose slit width just equals $R/2$. Under those conditions, the observed signal-to-noise, denoted $q$, is the empirical quantity of interest to us here. The signal-to-noise ratio $q$ has, in fact, been measured for the Lunar Orbiter film system (Elle,

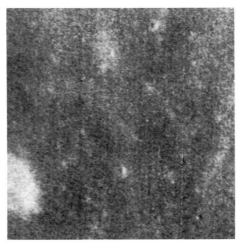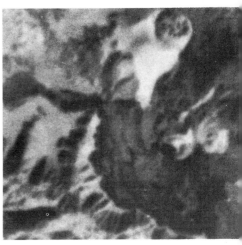

Each of the three frames above contains about the same amount of photographic information. Yet only the first one is easily recognizable as a telescopic view of Mars of about 100 kilometers surface resolution. Figure A.4 illustrates that the middle frame is one piece of a Ranger picture of the Moon with a surface resolution of about one kilometer. Figure A.5 and A.6 show that the view at right above is a degraded version of an aerial photograph of desert volcanic terrain.

*et al.*, 1967) and "SNRS considerably in excess of one were nevertheless obtained." A value of $q$ of about 1.4 is indicated by Figure 5 in Elle. Brock (1969, pp. 65), on the other hand, comments that ". . . the modulation of density across the target image is found to be approximately equal to the granularity calculated for an aperture having the area of one bar of the target. Thus the signal-to-noise ratio is on the order of one." Finally, Fulton (Personal communication) finds in an unpublished survey $q$ to be about 2.5. Thus $q$ seemingly lies between 1 and 2.5. We will use the value of 1.5 here, presuming that the uncertainty in its use does not exceed a factor of 2. Therefore,

$$\frac{V_S}{V_{N_R}} = q \qquad (15)$$

But what is the corresponding input $S/N$? We know that the observed input signal corresponded to a particular contrast $c$. If we represent the responsivity $\rho$ and the appropriate *MTF* corresponding to the limiting resolution $R$ as $Z(R_\rho)$, then

$$V_{S_R} = [i_{max} - i_{min}]Z(R_\rho)$$

$$= [i_{max} - i_{min}] \left[\frac{i_{max} + i_{min}}{i_{max} + i_{min}}\right] Z(R_\rho) \quad (16)$$

Since $i_{max} + i_{min} = 2i$,

$$V_{S_R} = 2i\left[\frac{c-1}{c+1}\right]Z(R_\rho) \qquad (17)$$

and

$$V_{N_R} = \delta i \ Z(R_\rho) \qquad (18)$$

therefore,

$$\frac{i}{\delta i_R} = \frac{q}{2}\left[\frac{c+1}{c-1}\right] \qquad (19)$$

## A.5 THE FUNDAMENTAL SIGNIFICANCE OF A PRIORI IGNORANCE

So far, we have followed the notion of a photograph as a link in a communication system which relays the "view from space" to the ground-based observer as faithfully as possible, or with intentional enhancement in some cases. Yet, the ultimate purpose of all this is to convey information* to the photo interpreter. He must recognize familiar features and detect and investigate unfamiliar ones. Thus it is necessary to relate the technical parameters of resolution and total

---

* We have tried to distinguish between communicated *data* with which to reconstruct photographs, measured in units of bits, and communicated *information* which also can be measured in bits. A photograph which is entirely incomprehensible to an observer because of his lack of familiarity with the scene viewed *contains no information,* regardless of how many bits were required to communicate it.

129

See caption for Figure A.3 for further details.

number of bits to the more subjective notion of interpretability. The key to this relationship is the extent to which prior knowledge of, or familiarity with, the surface features included in the scene is available. In fact, the basic relationship can be stated as follows: The interpretability of the physical features displayed in a photograph depends on (1) the total information content, and (2) the observer's familiarity with or prior knowledge of those surface features revealed at

**A.5   Progressive degradation of aerial view of volcanic terrain**

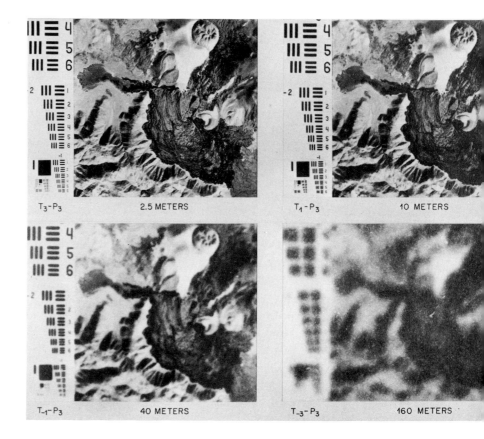

Four different resolutions of the same scene are presented above. Original aerial photograph was taken over the Coso domes area near Little Lake, California.

the surface resolution of that photograph. There is no independent significance to ground resolution per se as is illustrated in Figs. A.3, A.4, A.5, and A.6. Thus in the early phases of planetary exploration, ground resolution may be regarded as a dependent variable, to be adjusted in increasing steps from a scale comparable to prior photography down to some limit determined by the total information capability of the mission as well as by technical limitations to high-resolution photography.

We are led to conclude, therefore, that any space photography endeavor must be conceived, carried out, and ultimately interpreted in terms of what a priori information also exists, especially photographic. The information actually communicated by means of a space photography system depends entirely on how many new features can be recognized by the interpreter. His capability to recognize new features depends on his a priori knowledge concerning similar or related features. Thus the amount and nature of a priori knowledge should be the principal guide as to how the communicated bits are to be

expended, especially regarding the balancing of geographic coverage vs. ground resolution within whatever total information return limitation is placed by spacecraft and mission considerations.

## REFERENCES

Brock, G. C., Microimage Quality, Chapter 3, *Photographic Consideration for Aerospace*, Itek Corporation, Lexington, Mass., 1965.

Brock, G. C., Resolution and Micro Image Quality in Photographic and Other Systems, Appendix F, *Useful Applications of Earth-Oriented Satellites*, 6, National Academy of Sciences, Washington, D.C., 1969.

Elle, B. L., C. S. Heinmiller, P. J. Fromme, and A. E. Neumer, "The Lunar Orbiter Photographic System," *Journal of the SMPTE*, Vol. 76, No. 8, August 1967.

Fulton, James, CBS Laboratories, personal communication, 1969.

Smith, F. Dow, "Optical Image Evaluation and the Transfer Function," *Applied Optics*, Vol. 2, No. 4, April 1963.

**A.6   Geologic and topographic explanations**

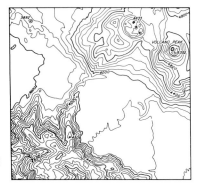 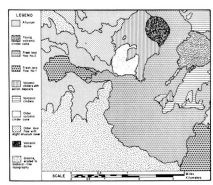

Explanatory maps for scene degraded in Figure A.5.

# APPENDIX
## B. Space camera design

### B.1 CAMERA TYPES

The design of cameras for use in space has emerged from a highly competent industry geared to the production of aerial cameras and electronic equipment. Thus, techniques and design concepts developed and tested in aerial cameras were immediately available for application to space systems. The decade of the fifties witnessed tremendous improvements in aerial photography brought about by the exploitation of a new platform—the high-flying jet aircraft that is relatively free from vibration—and the development of greatly improved photographic films as well as the design and construction of high-performance photographic lenses.

Since spacecraft are virtually vibration free, they are even better platforms for photography than are jet aircraft. Consequently, excellent performance can be expected by incorporating proper design. Spacecraft stabilized by horizon sensors, which view the Earth as do aircraft, can readily incorporate aircraft camera techniques. Other spacecraft, like the Mariners, are inertially stabilized. However, some spacecraft are spin-stabilized, and new camera designs are required to take pictures efficiently from rotating platforms; these are unique to space applications.

All cameras contain three basic elements; a lens to focus the image, a shutter to expose the image, and a sensor to record the image. The sensor might be typically photographic film or a television vidicon tube. Cameras are commonly configured as frame, strip, or panoramic type and the preferred choice depends upon the particular application. A summary descriptive of each type is presented next. Although the frame camera has been the most popular, the drive during the last decade for improved resolution has led to the use of strip and panoramic designs.

The *frame camera* is designed so that the lens focuses the image to be recorded on a frame of film or on the photosensitive surface of a television tube to be exposed by a single action of the shutter. The shutter may be a between-the-lens type or a louver type, both of which expose the entire frame simultaneously, or it might be a focal-plane type which exposes the frame by moving a curtain with an open slit in front of the film (see Fig. B.1). All hand-held cameras, ordinary motion picture cameras, and commercial television cameras are of the frame type, as are the cameras carried by astronauts in the Mercury, Gemini, and Apollo flights, and the Konvas motion picture camera carried by the Soviet cosmonauts.

Aerial cameras are usually programmed so that overlapping frames form a strip on the ground (see Fig. B.2), although isolated pictures may also be taken. This overlap is designed to give continuous coverage and a stereo view of the surface. The Lunar Orbiter and Mariner Mars spacecraft have been programmed in this way.

The image of the surface of a planet on the focal plane will move during exposure when photographed from a moving aircraft or space-craft. The amount of ground motion projected in the image plane or smear which can be tolerated depends upon the system resolution and is normally determined by the mission objectives and the camera design. When very high resolution is needed, the optical image or film is moved during the time of exposure to compensate for the smear. Image motion compensation (IMC) is

## B.1 The frame camera and the strip camera

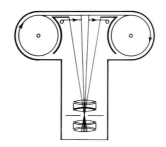

SCHEMATIC OF FRAME CAMERA

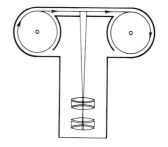

SCHEMATIC OF STRIP CAMERA

included on all modern aerial cameras and has been used on some space cameras, such as the Lunar Orbiter system.

The *strip camera* (Sonne, 1943), was designed to avoid the high-cycling rates necessary when using a frame camera in an aircraft flying at high speed and low altitude. In this design there is no operating shutter which opens and closes for film exposure. The lens of the strip camera focuses the image of the ground on the film just behind a fixed slit which acts as the shutter. The movement of the film is synchronized with that of the image so that no relative motion occurs during exposure. The amount of exposure is controlled by the width of the slit and by the velocity of the film (see Fig. B.1). The film coverage corresponds to a strip on the ground (see Fig. B.2). This type of camera has been used extensively to obtain high-resolution pictures from aircraft and no doubt will be popular from spacecraft when very high resolution is needed. It should be noted that this camera can operate only from a moving platform.

The frame camera (left) is the most popular design for use on the ground as well as from airplanes and spacecraft. In operation a frame is exposed by a single action of the shutter (focal plane or between-the-lens). The strip camera (right) uses the forward motion of the airplane or spacecraft to expose the film. The ground is imaged by the lens on the film through a slit, and the film is moved at a rate to compensate for the ground motion. The amount of exposure is controlled by the width of the slit and the velocity of the film.

The *panoramic camera* has been used from fixed installations on the ground for over one hundred years; however, its adaptation to aircraft use is fairly new. The panoramic camera scans the ground normal to the line of flight, while the strip camera scans the ground parallel to the line of flight (see Fig. B.2). The scan rate is adjusted

The strip camera photographs a single continuous strip on the ground; stereoscopic coverage can be obtained by tilting the camera forward for pictures and then later tilting it backward to photograph the same area on the ground. The frame camera can take a continuous ground strip by programming successive frames so that they overlap. The panoramic camera will take wide-angle pictures which can be programmed to overlap to obtain continuous coverage.

## B.2 Ground tracks of the strip, frame, and panoramic cameras

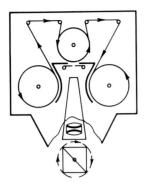

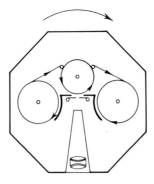

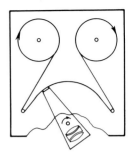

SCHEMATIC OF PRISM SCANNING
PANORAMIC CAMERA

SCHEMATIC OF SPINNING
PANORAMIC CAMERA

SCHEMATIC OF DIRECT SCANNING
PANORAMIC CAMERA

Different designs for panoramic cameras can be characterized by their method of scanning and whether the film is moving or fixed during exposure. The direct scanning camera (right) uses a rocking or spinning lens to perform the scan with the film fixed during exposure. The spinning panoramic or spin pan camera (center) operates from a spin stabilized spacecraft and uses the spin of the spacecraft to perform the scan. During exposure the film is moved past a slit at the proper velocity to compensate for the movement of the image of the ground. The mirror or prism scanning panoramic camera (left) uses a rotating mirror or prism to perform the scan and during exposure the film movement past a slit is synchronized with the scan.

to provide the desired overlap between exposures, as is commonly done with frame cameras. The first airborne panoramic test was made in 1949 by mounting a strip camera 90 deg to its normal position and rotating the entire camera to sweep a wide area across the flight line. The film was driven by the slit at a velocity sufficient to compensate for the rotation of the camera.

The popularity of the panoramic camera has increased rapidly during the last ten years and in many instances has replaced the frame camera. The reason lies in new improved designs exploiting the use of high-definition, narrow-angle lenses to achieve high resolution with wide-angle coverage. Many mechanical approaches have been proposed to perform the scan, and some have been built and tested. Basically these designs may be divided into three categories: the direct scanning camera, the rotating mirror panoramic camera, and the spinning panoramic camera.

The *direct scanning panoramic camera* uses

a rocking or rotating lens to perform the scan while the film is held firmly against the curved platten (see Fig. B.3). The lens is mounted in the scanning arm, which moves about the rear nodal point of the lens; the scan arm also contains a slit which acts as a shutter, exposing the film as the slit moves across the film. One advantage of this design is that the film does not move during exposure, thereby avoiding synchronization problems. Image motion compensation is normally performed by sliding the lens and scan arm along the camera's rotation axis as it scans. The amount of movement is controlled by a cam whose shape depends upon vehicle velocity, altitude, camera scan rate, and focal length.

The *rotating mirror panoramic camera* normally uses a spinning mirror or dove prism to scan and direct the image into a fixed lens system. The lens focuses the image on the film, which moves at a rate which compensates for the mirror scan. A fixed slit is placed in front of the film to act as a shutter (see Fig. B.3). The exposure is determined by the width of the slit and by the velocity of the film.

The *spinning panoramic camera* (Davies, 1964), sometimes called the spin pan camera, is designed for use in a spin-stabilized spacecraft. It thus differs from the direct scanning and rotating mirror camera types, which operate from either aircraft or planet-oriented spacecraft. This design uses the spin of the spacecraft to perform the scan of the lens. The lens focuses the image on the film, which is moving at the necessary velocity to compensate for the spin of the spacecraft (see Fig. B.3). The exposure is made

134

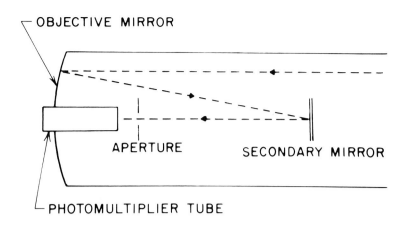

OBJECTIVE MIRROR

APERTURE          SECONDARY MIRROR

PHOTOMULTIPLIER TUBE

The line scan camera builds up a picture by focusing on one point at a time. Operating from a spinning spacecraft, the point sweeps out line after line. The forward velocity of the spacecraft will stagger the lines so that a two-dimensional picture is built up.

through a slit placed in front of the film, and the amount of exposure is determined by the width of the slit and by the velocity of the moving film.

The *line scan camera* is the simplest of all types, consisting of an optical system which focuses a point image on a photomultiplier. The output of the photomultiplier is used to build up a line picture as the scan is made by a mirror or by the spinning spacecraft. Successive scan lines are assured by the movement of the spacecraft along its trajectory, as with Pioneer F, or by the gradual tipping of the lens in the spacecraft, as with the ATS Earth pictures (see Fig. B.4).

Frame cameras and most panoramic types can operate from stationary platforms, and when used from aircraft or spacecraft image motion during exposure must be expected. Of course, with high shutter speeds or low vehicle velocities the smear can be ignored; however, when the blur is excessive it is necessary to compensate for the motion. On the other hand the strip type camera requires this motion to control the exposure and so can operate only from moving aircraft or spacecraft. The spinning panoramic and line scan cameras must operate from a spinning platform and so are true space cameras which can be used only on spin stabilized spacecraft. Thus the optimum choice of camera type is closely tied to the spacecraft design and mission performance objectives.

## B.2   THE EXPERIENCE SO FAR

Perhaps the most popular camera for use in space is the vidicon television camera; its simplicity of design, light weight, and reliability

of operation have contributed to its reputation. This frame type camera enjoyed early success in the TIROS program and since then has been used on many earth orbital, lunar, and planetary programs; it has been used as the primary imaging system and also on other occasions as an instrument to monitor the performance of man and devices. Of course, the television camera is designed to transmit images over a radio communication link so is a natural first choice for this type of operation.

Another principal mode of operation involves the physical return and recovery of a capsule from the spacecraft; all manned missions are of this type as are many unmanned. Photographic film is the most popular sensor for picture taking in these missions; camera types vary but the conventional frame camera is the most common. The format of these cameras is larger than that of the television cameras and the resolution is greater so each photographic film picture contains many more information bits than a vidicon picture. Film is a very efficient storage medium for pictorial data. Photographic film with onboard processing and readout to transmit the pictures to the ground for reassembly has been used on a number of missions, such as Luna 3, Zond 3, Luna 12, the Lunar Orbiters, and more.

Table B.1 gives the characteristics of a large number of cameras which have been used in

135

**Table B.1    Space cameras—flown and planned**

| Program | Camera type | Sensor | Storage | Lens Focal length mm | Lens Focal ratio F/No | Format cm × cm | Data return |
|---------|-------------|--------|---------|----------------------|-----------------------|----------------|-------------|
| Mercury-Hasselblad | Frame | Film | Film | 80 | 2.8 | 6 × 6 | Recovery |
| Gemini-Hasselblad | Frame | Film | Film | 80 | 2.8 | 6 × 6 | Recovery |
| | | | | 80 | 2.8 | | |
| Apollo-Hasselblad | Frame | Film | Film | 250 | 5.6 | 6 × 6 | Recovery |
| Hycon | Frame | Film | Film | 450 | 4 | 11 × 11 | Recovery |
| Itek | Rotating mirror panoramic | Film | Film | 610 | 3.5 | 11 × 113 | Recovery |
| Fairchild | Frame | Film | Film | 75 | 4.5 | 11 × 11 | Recovery |
| Vostok-Konvas (Vashkod) | Frame | Film | Film | 50 | 2 | 1.6 × 2.2 | Recovery |
| | | | | 75 | 3.5 | | |
| | | | | 135 | 2.8 | | |
| Zond 5, 6, 7 | Frame | Film | Film | 400 | 6.3 | 13 × 18 | Recovery |
| Lunar Orbiter | Frame | Film | Film | 80 | 5.6 | 5.5 × 6.5 | Readout |
| | | | | 610 | 5.6 | 5.5 × 21.9 | |
| Ranger | Frame | Vidicon | None | 76 | 2 | 1.14 × 1.14 | Readout |
| | | | | 25 | 0.95 | 1.14 × 1.14 | |
| | | | | 76 | 2 | .28 × .28 | |
| | | | | 25 | 0.95 | .28 × .28 | |
| Mariner 4 | Frame | Vidicon | Tape recorder | 305 | 8 | .56 × .56 | Readout |
| Mariner 6 & 7 (Mars '71 orbiter) | Frame | Vidicon | Tape recorder | 50 | 5.6 | .96 × 1.25 | Readout |
| | | | | 508 | 2.5 | .96 × 1.25 | |
| Luna 3 | Frame | Film | Film | 200 | 5.6 | 2.5 × 2.5 approx | Readout |
| | | | | 500 | 9.5 | 2.5 × 2.5 approx | |
| Zond 3 | Frame | Film | Film | 106.4 | 8 | 2.2 × 2.3 approx | Readout |
| Luna 12 | Frame | Film | Film | | | | Readout |

space or are planned for use in space. Included are a variety of camera types and sensors. Table B.2 gives the characteristics of a few cameras contained in proposals which received careful study and had some special significance.

The STL-Lunar Orbiter study is considered important because it demonstrated that the orbiter mission could be accomplished using the Atlas-Agena booster and need not be delayed until the Atlas-Centaur booster became available. When the capabilities of a lightweight spacecraft were recognized, a competition was held to select a contractor for the mission. Thus began the very successful Lunar Orbiter Program.

For some time there has been work on electro-optical cameras which store images on dielectric tape. The objective of these designs is to combine the best features of film and the television cameras; thus the high resolution image is stored on tape to be read out later, after which the tape can be reused. The operation of the camera can resemble that of a vidicon in which the image is focused on a photoconductor (perhaps antimony trisulfide) coating on the tape or that of an image orthicon in which the image is focused on a photoemissive cathods which ejects electrons in proportion to the intensity of light. These electrons strike the electrostatic tape, changing its conductivity locally. The RCA dielectric tape camera, developed for the Nimbus spacecraft, is the only system of this type which has been built and tested. A new and more efficiently designed tape camera was contained in the Davies/CBS camera proposal for the Pioneer F & G missions. Although these cameras may be considered complex, requiring the dielectric tape

| Program | Camera type | Sensor | Storage | Lens | | Format cm × cm | Data return |
|---|---|---|---|---|---|---|---|
| | | | | Focal length mm | Focal ratio F/No | | |
| Tiros | Frame | Vidicon | Tape recorder | 5.4 | 1.5 | .62 × .62 | Readout |
| | | | | 5.8 | 1.8 | .62 × .62 | |
| | | | | 40 | 1.8 | .62 × .62 | |
| Nimbus | Frame | Vidicon | Tape recorder | 17 | 4 | 1.12 × 1.12 | Readout |
| | | | | 5.8 | 1.8 | 1.12 × 1.12 | |
| ESSA | Frame | Vidicon | Tape recorder | 5.8 | 1.8 | 1.12 × 1.12 | Readout |
| Meteor (Cosmos 122, 144, 156, 184, 226) | Frame | Vidicon | Tape recorder | | | | Readout |
| Molniya | Frame | Vidicon | | | | | Readout |
| Dodge | Frame | Vidicon | | | | | Readout |
| ERTS | Frame | Vidicon | Tape recorder | 126 | 2.8 | 2.5 × 2.5 | Readout |
| Nimbus DTC* | Mirror scanning panoramic | Dielectric tape | Dielectric tape | 125 | 2.5 | 1.9 × 23.2 | Readout |
| Surveyor | Frame | Vidicon | | 25 | 4 | | Readout |
| | | | | 100 | 4 | | |
| Luna 9, 13 | Line scan | PM tube | None | 12.4 | 3 | | Readout |
| ATS-1 | Line scan | PM tube | None | 250 | 2 | | Readout |
| -3 | | | | 375 | 3 | | |
| Pioneer F, G | Line scan | PM tube | None | | | | Readout |
| Mariner Venus/ Mercury | Frame | Vidicon | Tape recorder | 1500 | 9 | .96 × 1.25 | Readout |

*The RCA dielectric tape camera was developed for the Nimbus B; it was built, tested, and space rated but never flown.

**Table B.2   Space cameras—proposals and studies**

| Program | Camera type | Sensor | Storage | Lens | | Format cm × cm | Data return |
|---|---|---|---|---|---|---|---|
| | | | | Focal length mm | Focal ratio F/No | | |
| STL-Lunar Orbiter Study | Spinning panoramic | Film | Film | 1500 | 6 | 6.0 × programmed | Readout |
| Davies/CBS- Pioneer Proposal | Spinning panoramic | Dielectric tape | Dielectric tape | 250 | 2.5 | 1.2 × programmed | Readout |
| Mariner Venus/ Mercury Film Study | Frame | Film | Film | 1000 | 6 | 2.4 × 2.4 | Readout |

to move inside the vacuum envelope, they can be lightweight and have tremendous storage so there may be a role for them in future missions.

The Mariner Venus/Mercury Imaging Science Team made a study comparing various vidicon television systems with a photographic film readout system (see Appendix C). This film system received a great deal of study by the team and by interested contractors who verified the feasibility and practicality of the design.

137

**Table B.3  Alternative combinations of sensor and data return**

| DATA RETURN | SENSOR | | |
|---|---|---|---|
| | Film | Vidicon | Photomultiplier |
| Recovery | Mercury<br>Gemini<br>Apollo<br>Vostok<br>Voshkod<br>Zond 5, 6, 7 | Not applicable | Not applicable |
| Real time | Not applicable | Dodge<br>Ranger<br>Surveyor<br>Molniya 1 | Explorer 6<br>ATS 1, 3<br>Pioneer F, G<br>Luna 9, 13 |
| Stored video | Lunar Orbiter<br>Nimbus DTC<br>STL Lunar Orbiter<br>  Study<br>Davies/CBS Pioneer<br>  Proposal<br>Mariner Venus/<br>  Mercury Film Study<br>Luna 3<br>Zond 3<br>Luna 12 | Tiros<br>Mariner 4<br>ESSA<br>Nimbus 1, 2<br>Mariner 6, 7<br>Mariner 1971<br>  Mars Orbiter<br>Mariner Venus/<br>  Mercury<br>Meteor | None<br><br><br><br><br><br><br><br>None |

Mission objectives and spacecraft characteristics go a long way toward defining camera sensors and camera types which should be considered in the planning phase of a new space program. To illustrate the influence of certain parameters Table B.3 was prepared, in which the various cameras of Tables B.1 and B.2 are ordered with regard to the sensor used (film, vidicon, or photomultiplier) and the method of data return (physical recovery, real time readout, or stored readout). Similarly, in Table B.4 these same cameras are listed by camera type (frame, strip, panoramic/line scan) and the method of spacecraft stabilization (inertial, planetary reference, and spin). In each category, the U.S. programs are listed first and the Soviet second. In the future, with more experience and confidence, it is likely that there will be more variety in camera designs as there will be increasing pressures for better pictorial data and the use of special cameras to obtain them most efficiently.

## REFERENCES

Davies, Merton E., U.S. Patent 3,143,048, August 4, 1964, Photographic Apparatus.

Jensen, Niels, *Optical and Photographic Reconnaissance Systems*, New York, John Wiley and Sons, 1968.

Magill, Arthur A., Chapter 4, Still Cameras, Vol. IV, *Applied Optics and Optical Engineering*, Editor, R. Kingslake, Academic Press, 1967.

Sewell, Eldon D., Chapter IV Aerial Cameras, Vol. I, *Manual of Photogrammetry*, American Society of Photogrammetry, 1965.

Sonne, Frederick T., U.S. Patent 2,307,646, January 5, 1943, Camera.

**Table B.4 Alternative combinations of spacecraft stabilization and camera type**

CAMERA TYPE

| STABILIZATION | Frame | Strip | Panoramic/line scan |
|---|---|---|---|
| 3 axis inertial | Mercury<br>Gemini<br>Apollo-Hasselblad<br>    -Hycon<br>    -Faiirchild<br>Ranger<br>Mariner 4<br>Mariner 6, 7<br>Mariner 1971 Orbiter<br>Mariner Venus/Mercury<br>Mariner Venus/Mercury<br>  Film Study | None<br><br><br><br><br><br><br><br><br><br>None | Apollo-Itek<br><br><br><br><br><br><br><br><br><br>None |
| | Vostok<br>Voshkod<br>Zond 5, 6, 7 | | |
| Planet oriented | Dodge<br>Lunar Orbiter<br>Nimbus 1, 2<br>Surveyor<br>Luna 3<br>Zond 3<br>Molniya 1<br>Luna 12<br>Meteor | None<br><br><br><br>None | Nimbus DTC<br><br><br><br>Luna 9, 13 |
| Spin | Tiros<br>ESSA<br><br><br><br><br><br>None | Not applicable | Explorer 6<br>ATS 1, 3<br>Pioneer F, G<br>STL Lunar Orbiter<br>  Study<br>Davies/CBS Pioneer<br>  Proposal<br>None |

139

# APPENDIX
## C. Performance criteria

The designer must always deal with alternatives in which cost and reliability must be balanced against performance. Thus criteria for imaging system performance must be developed in a form useful to both the scientist and the engineer. The next two sections develop such criteria and, in the last section, these results are applied in an illustrative example.

### C.1 A FIGURE OF MERIT FOR RESOLUTION

Achievable ground resolution is of paramount importance in photographic exploration. It is often necessary to compare alternative imaging systems in terms of their expected ground resolution. However, performance of television systems is often specified in terms of signal-to-noise per television line and an MTF curve, while that of photographic systems is often specified in terms of limiting bar-chart resolution at some specified contrast. Furthermore, photographs taken from the Earth, which necessarily supply the a priori background for early planetary missions, are normally evaluated in terms of the associated seeing disk or assigned a limiting ground resolution subjectively by an astronomer. How can this variety of imagery be placed on a common basis as far as ground resolution is concerned? Our proposal is simple: The most unbiased basis for evaluation of ground resolution is the relative amount of photographic information returned per unit surface area of the planet. And the information content of the various kinds of imagery can be estimated from the number of bits per unit surface area required to encode for transmission, without significant degradation, the scene intensities reconstructed from each kind of imagery. Specifically, we define the figure

of merit for resolution, FOM, as the number of bits required for a particular system to encode without significant degradation the intensity reconstructed from an image of a unit surface area of the target (expressed in square km).

Equation (A-6) forms the basis of expressing the FOM for a television system, except that we must relate flat-field (uniform input intensity) signal-to-noise to a $S/N$ which refers to some spatially varying intensity. This relationship can be obtained through (1) the system $MTF$ curve, which expresses the appropriate intensity attenuation factor as a function of television lines, and (2) the assumption that the noise among adjacent television lines is uncorrelated. Under these circumstances, the equivalent input signal-to-noise, $i/\delta i$, is related to the flat-field output $S/N$,

$$\frac{i}{\delta i} = (MTF)\left[\frac{K_0}{K_{MTF}}\right]\left[\frac{S}{N}\right]_F + 1 \quad (C-1)$$

where $MTF$ equals the attenuation coefficient for a particular space frequency, $K_{MTF}$ in television lines/mm corresponding to that $MTF$, and $K_0$ is the television lines/mm in the format, and $(S/N)_F$ is the flat-field signal-to-noise.

At this point, we come to the same kind of arbitrariness encountered with film systems when it is necessary to specify a particular contrast at which limiting resolution will be specified. In both cases some form of weighted integral would seem indicated. Yet in both cases the relevant test data for such integrals are rarely available until *after* system choices have been made. Also, the use of integrals implies some kind of a priori weighting of the expected spatial and contrast content of the unknown scene and thus a bias as to the results. Hence, it is desirable in practice to choose a nominal *MTF* for evaluation of television systems, recognizing that some oversimplification is necessarily involved. In the present case we choose the *MTF* value of 0.2 as the highest spatial resolution point on the *MTF* curve which can be reliably estimated in advance from design data. Accordingly

$$\frac{i}{\delta i} = 0.2 \left[ \frac{K_0}{K_{0.2}} \right] \left[ \frac{S}{N} \right]_F + 1 \qquad \text{(C-2)}$$

and

$$\text{FOM}_{\text{TV}} = \frac{3Yf^2 K_{0.2}{}^2}{H^2} \log_{10} \left[ 1 + \frac{0.2\ K_0}{K_{0.2}} \left( \frac{S}{N} \right)_F \right] \qquad \text{(C-3)}$$

where $f$ is the effective focal length in mm and $H$ the range in kilometers to the target.

Similarly, Equation (A-20) yields

$$\text{FOM}_{\text{FILM}} = \frac{12R^2 Yf^2}{H^2} \log_{10} \left[ 1 + \frac{q(c+1)}{2(c-1)} \right] \qquad \text{(C-4)}$$

And, if we denote *S* to be the minimum detectable dimension of a surface feature of intrinsic contrast *C* in a ground-based planetary photograph,*

$$\text{FOM}_{\text{GB}} = \frac{12Y}{S^2} \log_{10} \left[ 1 + \frac{q(c+1)}{2(c-1)} \right] \qquad \text{(C-5)}$$

As an example, we will compare the number of bits per unit surface area which could be returned from a high-resolution Lunar Orbiter

camera system hypothetically placed in orbit about Mars with the number to be returned from the Mariner TV camera system to be deployed in 1971.

Combining Equations (C-3) and (C-4), and assuming both systems operate from a 1500 km closest approach

$$\frac{\text{FOM}_{\text{LO}}}{\text{FOM}_{\text{M'71}}} = \frac{4R_{\text{LO}}{}^2 f_{\text{LO}}{}^2 \log_{10} \left[ 1 + \frac{q(c+1)}{(c-1)} \right]}{K_{0.2}{}^2 f_{\text{M'71}}{}^2 \log_{10} \left[ 1 + \frac{0.2\ K_0}{K_{0.2}} \left( \frac{S}{N} \right)_F \right]} \qquad \text{(C-6)}$$

and, noting from Table C.1 and Fig. C.1 that

$$(S/N)_F = 75$$
$$f_{\text{LO}} = 610 \text{ mm}$$
$$f_{\text{M'71}} = 508 \text{ mm}$$
$$K_{0.2} = 44 \text{ mm}^{-1}$$
$$R_{\text{LO}} = 76 \text{ } \ell\rho/\text{mm at contrast of } 3{:}1$$
$$K_0 = 74 \text{ mm}^{-1}$$
$$q = 1.5$$

then,

$$\frac{\text{FOM}_{\text{LO}}}{\text{FOM}_{\text{M'71}}} = \frac{4(76)^2\ (610)^2\ \log_{10}\ [1 + 1.5]}{(44)^2\ (508)^2\ \log_{10} \left[ 1 + \frac{74}{30} \cdot \frac{75}{5} \right]}$$

$$= 5 \qquad \text{(C-7)}$$

Thus, even at a range of 1500 km, a Lunar Orbiter system would appear to yield half an order of magnitude greater information per unit surface area than will the kind of system to be flown around Mars by the United States in 1971.*

---

* Inasmuch as the minimum detectable pattern in such a case is not a repetitious one, the appropriate value of q may be higher than for the film test discussed earlier. However, we will continue to use the value 1.5 in the absence of published data to the contrary.

* Keene (Private communication) indicates that actual flight performance of the Lunar Orbiter system was higher than the test value of 76 1p/mm quoted here, and, in fact, gave at least 100 1p/mm at 3:1 contrast. Use of this performance number would increase the apparent advantage over the television system.

**Table C.1 (a)    Mariner Mars television cameras**

| Optics | Mariner 4 (1965) | Mariner 6 & 7 (1969) | | 1971 Orbiter | |
| --- | --- | --- | --- | --- | --- |
| | | Camera "A" | Camera "B" | Camera "A" | Camera "B" |
| Type | 2 mirror cassegrain | 6 element lens | 3 element catadioptric | 6 element lens | 3 element catadioptric |
| Focal length | 30.5 cm | 5 cm | 50.8 cm | 5 cm | 50.8 cm |
| f/no | f/8 | f/5.6 | f/2.5 | f/4 | f/2.5 |
| Shutter | | | | | |
| Type | Intermittent rotary | Intermittent rotary | Twin solenoid two blade | Twin solenoid two blade | Twin solenoid two blade |
| Exposure range | .08–.20 sec | .06–.12 sec | .006–.012 sec | .024–.192 sec | .006–.048 sec |
| Filters | Green, orange | Green, orange, blue | Minus blue | UV, blue, green, orange, 3 polarizers, minus blue | Minus blue |
| Sensor | | | | | |
| Type | Slow scan vidicon | Slow scan vidicon | Slow scan vidicon | Slow scan vidicon | Slow scan vidicon |
| Format (mm) | $5.6 \times 5.6$ | $9.6 \times 12.5$ | $9.6 \times 12.5$ | $9.6 \times 12.5$ | $9.6 \times 12.5$ |
| Format (pixels) | $200 \times 200$ | $704 \times 945$ | $704 \times 945$ | $700 \times 832$ | $700 \times 832$ |
| Field of view (degrees) | $1.05° \times 1.05°$ | $11° \times 14°$ | $1.1° \times 1.4°$ | $11° \times 14°$ | $1.1° \times 1.4°$ |
| Frame time (sec) | 24 | 42 | 42 | 84 | 84 |
| Storage | | | | | |
| Type | Magnetic tape digital | Magnetic tape mixed analog and digital | | Magnetic tape digital | |
| A/D encoding | 6 bits | 8 bits | | 9 bits | |
| Total storage | $5 \times 10^6$ bits | $1.8 \times 10^8$ bits | | $1.8 \times 10^8$ bits | |

Furthermore, the Lunar Orbiter type system has a built-in image motion compensation system (which also could be provided for a television system) permitting it to be operated much closer to the surface than can the Mariner '71 system, which is virtually smear limited at about 1000 km. Hence, it is clear that the basic space-proven technology already exists which could make possible a dramatic breakthrough in Mars surface resolution, assuming the weight, power, and cost increments were deemed to be warranted.

Another kind of comparison which must be carried out frequently involves the comparison of space probe photography with good ground-based planetary photographs. We ask the very real operational question: At what distance away from Mars would spacecraft photography of the whole disk have an information content approximately equal to that of the very best ground-based photographs?[**] Assuming the spacecraft to be of the Mariner Mars '69 or '71 type, and equating Equations (C-4) and (C-5),

$$\frac{3Yf^2K_{0.2}^2}{H^2} \log_{10}\left[1 + \frac{0.2K_0}{K_{0.2}}\left(\frac{S}{N}\right)_F\right]$$
$$= \frac{12Y}{S^2} \log_{10}\left[1 + \frac{q(c+1)}{2(c-1)}\right] \qquad (C–8)$$

and

$$H^2 = \frac{S^2f^2K_{0.2}^2O}{4} \frac{\log_{10}\dfrac{0.2K_0}{1+K_{0.2}}\left(\dfrac{S}{N}\right)_F}{\log_{10}\left[1 + \dfrac{q(c+1)}{2(c-1)}\right]} \qquad (C–9)$$

Taking the very best ground-based resolution to be 100 km for a feature of 2:1 contrast, and

[**] It would be desirable to begin even earlier in order to have the benefit of continuous, uniform coverage.

using the previous values for the '71 system, we find $H = 1.8 \times 10^6$ km. This corresponds to about three days before encounter in the case of 1969 or 1971 and is in approximate agreement with the actual results from the far-encounter pictures acquired by Mariners 6 and 7. Analogous calculations for Jupiter and Mercury for the same kind of camera system indicate equivalent information content at about seventeen days ($12 \times 10^6$ km)

out from Jupiter and at a little more than two days ($2.2 \times 10^6$ km) from Mercury.*

* Assuming a best ground-based resolution of 250 km at 10 percent contrast for Mercury and an average night-to-night ground-based resolution of 1200 km at 20 percent contrast for Jupiter, the appropriate value for a temporally varying object. We are indebted to B. A. Smith for these numbers. Encounters in 1973 were assumed with approach speeds around 13 km/sec for Mercury and 8 km/sec for Jupiter.

**Table C.1 (b)  Comparison of Mariner Mars '71 and Lunar Orbiter\* camera systems**

|  | Lunar Orbiter | | Mariner Mars '71 | |
|---|---|---|---|---|
| System weight | 147 lb | | 68.3 lb | |
| Camera types | Two frames | | Two frames | |
| Sensor | Photographic films | | Two vidicons | |
| Lens focal lenth | 80 mm | 610 mm | 50 mm | 508 mm |
| focal ratio | F/5.6 | F/5.6 | F/4 | F/2.5 |
| Format | 38° × 44° | 5.2° × 20.4° | 11° × 14° | 1.1° × 1.4° |
|  | 55 × 65 mm | 55 × 219 mm | 9.6 × 12.5 mm | 9.6 × 12.5 mm |
| Read-out pixels/line | 9880 | 33,288 | 832 | 832 |
| lines/frame | 8360 | 8,360 | 700 | 700 |
| Pixels per frame | $8.3 \times 10^7$ | $2.8 \times 10^8$ | $5.8 \times 10^5$ | $5.8 \times 10^5$ |
| System resolution parameters | Limiting resolution of 76 lp/mm at 3:1 contrast and S/N = 1.5 | | Flatfield signal-to-noise ~ 75/1, 20% response at 44 TV lines/mm | |
| Projected frame size (from 1500 km) |  |  |  |  |
| Along flight | 1035 km | 135 km | 285 km | 28.5 km |
| Cross flight | 1215 km | 540 km | 375 km | 37.5 km |
| Projected pixel size (from 1500 km) | 108 m | 16 m | 420 m | 42 m |
| Periapsis smear at 3.5 km/sec | Compensated | | 21 m @ 6 ms | 10 m @ 3 ms |
|  |  |  | 168 m @ 48 ms | 89 m @ 24 ms |

* The pixel size and line widths shown here refer to the design performance of 76 lp/mm and 152 scan lines/mm. The actual lunar orbiter system used 278 scan lines/mm and is reported to have exceeded 100 lp/mm at 3:1 contrast.

## C.1 Resolution of space camera systems

The modulation transfer curves for three camera systems discussed in this book are presented above. The modulation transfer function is defined in Appendix A.

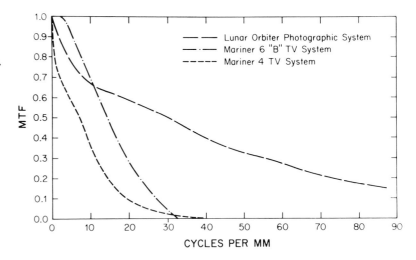

The FOM we have discussed in this section is based solely on relative surface information communicated; it does not distinguish between high signal-to-noise (and large pixels) and lower signal-to-noise (and smaller pixels). Should a priori information or attitudes be available such that the desired signal-to-noise per pixel could be stated in advance, then a more responsive FOM could be developed based upon the relative surface area imaged by each system at the specified signal-to-noise. This procedure affords a way of taking into account more fully the differing *MTF* curves and noise distributions of various candidate systems.

## C.2 COVERAGE, RESOLUTION, AND TOTAL DATA RETURN

The preceding discussion of a figure of merit for resolution provides one basis for comparison of different camera-systems. However, as was discussed in Appendix A, resolution is only one aspect of the problem. Total data return and geographic coverage must also be considered. In addition, usually there are secondary, but still very important, considerations such as photometric and geometric usefulness of the data, spectral coverage, and so on. And cost and technical readiness must be balanced against all of these factors. Thus, in practice the total scientific results likely to result from the entire imaging *system* must be assessed for each alternative

option to provide a basis for a cost/effectiveness choice. A useful step in developing such a comparison is the tabulation of surface coverage or total data return as a function of surface resolution. The development and presentation of such tabulations is the subject of this section. In the next section (C.3) these results are applied to the actual problem of finding an optimum technical solution for a space photography problem within real constraints.

To tabulate surface coverage versus surface resolution for a particular imaging experiment, the first step is to identify the size of a characteristic resolution element in individual frames, using consistent criteria such as discussed in the previous section. For example, in the Mariner 4 television pictures, the 10 percent response point on the *MTF* curves corresponded to 20 line pairs/mm (see Fig. C.1). This might be chosen as the resolution element of interest. Since the format of the vidicon tube used was about 5 mm $\times$ 5 mm, it is evident that there are about $100 \times 100$ resolution elements per frame, or a total of about $10^4$. The focal length of the camera was 30 cm (Table C.1). Thus at a range of 10,000 km, for instance, each resolution element corresponded to a unit of 1.6 km $\times$ 1.6 km on edge. The frame would be described as containing $10^4$ units of 1.6 km resolution. A similar procedure is then carried out for each frame in the series (except that areas of overlap in adjacent pictures

## C.2    Photographic exploration of Mars—a quantitative view

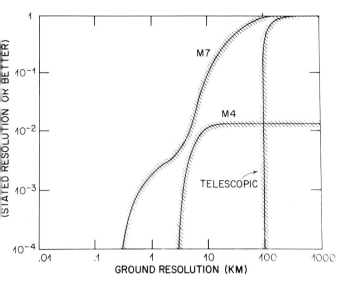

This diagram tabulates the total surface coverage photographed at a resolution better than the given value shown. Hence, for Mariner 7 (indicated as M7), about 10 percent of the surface was photographed at a resolution of 10 kilometers or better. Mariner 4 and ground-based telescopic information are also plotted. Additional information is available in the text.

## C.3    Photographic exploration of the Moon—a quantitative view

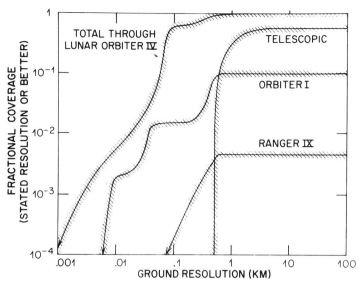

A similar diagram as C.2. Note than Ranger IX and Orbiter I extend in resolution off the bottom of the chart to the intersection with about one-meter resolution.

are only counted once). Also, a reduction must be made for black sky away from the planet in some views or for scenes viewed beyond the terminator (on the nighttime surface). Furthermore, in views where curvature of the planetary surface is evident, degraded surface resolution must be associated with the appropriate optical resolution elements within each frame. The resultant sets of numbers of resolution elements versus surface resolution are then arranged in cumulative form and converted to fractional area by normalization to the total planetary area. These normalized cumulative data are then plotted on log/log paper for convenient viewing, as is illustrated in Fig. C.2. There the Mariner 4 coverage is contrasted with that of Mariner 7 and previous Earth-based telescopic coverage. This presentation tends to put in more quantitative form the necessity of adjusting each successive step in resolution to ensure an orderly transition from known to unknown surface resolution (as discussed in Appendix A). It can be seen from Fig. C.2 that the step from telescopic data to

Mariner 4 was too abrupt because intermediate resolution full-disc coverage could not be obtained. The Mariner 7 curve, however, shows a much smoother transition, except between 1 and 10 km, where the remaining gap between far- and near-encounter sequences is still evident.

Figure C.3 provides similar curves for key lunar photographic missions preceding Apollo and illustrates dramatically how the photographic exploration of that body proceeded during the 1960s. Finally, in Fig. C.4, Mariner 7 is compared with the lunar history, providing a semi-quantitative description of the present state of Martian photographic exploration in terms of our lunar experience. Thus, geographic coverage versus surface resolution can be a valuable parameter of photographic exploration and affords a way to compare a proposed mission with previous ones.

However, there is an alternative kind of presentation which can be especially useful in considering alternative imaging approaches to the same mission. That form is based on the tabulation of total communicated bits correspond-

145

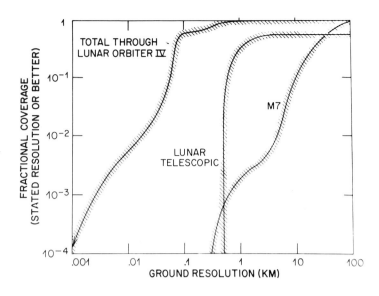

It can be seen from this diagram that even Mariner 7 has hardly brought Mars beyond the pre-Space Age information level of the Moon.

ing to a given surface resolution, rather than fractional surface coverage. The procedure to obtain the cumulative data is similar to that for fractional surface coverage except that the total bits for each frame are associated with the resolution of that frame. In this compilation, overlapping frames are included (but not black sky). Hence, intentional overlap to obtain color coverage or stereo is more properly accounted for. In any case, since total bits communicated to Earth are often the most significant limitation to space photography, the manner in which those bits are expended generally reflects very careful evaluation by the designers.

Figure C.5 presents the Mariner 4 and Mariner 7 picture data in the form of total bits recovered at stated resolution or better versus ground resolution.

## C.3    EVALUATION OF ALTERNATIVE SYSTEMS

In Section C.1, a basis was established for comparison of differing imaging systems in terms of ground resolution. In Section C.2 a means was then developed to display the coverage/resolution relationship and total data/resolution relationship. It is the purpose of this section to illustrate how these relationships can be used

## C.4    Comparison of the photographic exploration of Mars and Moon

in actual practice to provide the basis for instrument choice in an actual mission.

In September 1969, NASA selected an advisory team of seven scientists to aid in the design of the imaging experiment on a spacecraft intended to be launched toward Venus in late 1973. On arrival at Venus in early February of 1974, the spacecraft would be targeted so as to be deflected toward an encounter with Mercury about seven weeks later. This mission, now referred to as the Mariner Venus/Mercury '73 mission, is to be carried out at minimum cost and to use previously designed components, including scientific instruments. The Imaging Advisory Team analyzed various instrumentation alternatives in terms of coverage, total data return, and surface resolution and presented these in several reports. This process materially affected the final choice for the MVM'73 mission and provides a precedent for future mission design exercises. Thus, it is useful to summarize the MVM'73 analyses here. However, we will only consider a few of the options that were actually considered in that study and we shall use consistent and finally corrected input data.

The steps in the MVM'73 analysis were six:

1.) Definition of scientific objectives for imaging.

2.) Identification of mission constraints and nominal mission profiles.

3.) Enumeration and description of instrument alternatives.

4.) Determination of picture sequences for each alternative to maximize scientific return within mission constraints.

5.) Tabulation and presentation of coverage, total data return, and surface resolution curves

**C.5 An alternative diagram**

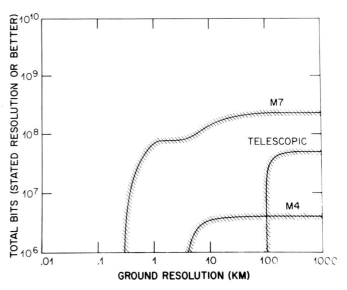

In this diagram the total bits of picture data are tabulated as a function of ground resolution instead of surface coverage. This kind of diagram better takes into account stereo, color overlap, and so on.

(as in Section C.2) for each instrument alternative.

6.) Comparison of alternatives in terms of: (1) scientific value, as indicated by the above curves; (2) cost, and; (3) readiness.
These steps are, in fact, part of an iterative process, especially as items 2 and 3 tend to change during a design study.

The scientific objectives for imaging of Mercury and Venus are discussed extensively elsewhere. The most significant impact of those objectives upon Mariner camera systems (which have been developed to photograph Mars) are:

1.) Extension of the spectral coverage into the ultraviolet to study the mysterious ultraviolet features of Venus.

2.) Increase of the angular resolution of the camera to overcome the greater range to the surface of Mercury inherent in any flyby trajectory which passes through the solar occultation zone (as is desirable to determine Mercury's interaction with the solar wind).

3.) Provision of greater total picture data from Mercury than previous Mars missions in order to bridge the gap from the much worse a priori data to geologically useful surface resolutions (1 kilometer or better). Inasmuch as there are no plans or prospects for any follow-on flights to Mercury by the United States, the picture data returned by the single MVM'73 spacecraft probably will not be added to for a decade or more. Thus, this one mission must attempt as complete a transition as possble from our present state of ignorance to whatever maximum resolution can be achieved.

The most demanding of the above three factors is the last. Due to the high relative velocity of

the spacecraft relative to Mercury ($\sim 11$ km per second or about twice that of a Mars flyby) and the unfavorable viewing geometry, the observation time is brief. Thus, if a conventional Mariner-type television system is used, the number of high resolution pictures is limited to a single tape load. This has also been the case for Mariners 4, 6, and 7, since during the playback time the spacecraft moves out of high resolution range. The instrument option "TV-12" described in the accompanying table, is a system in which only minor changes are incorporated from the Mariner cameras to be used in the Mars orbiters of 1971 (and which, in turn, represent only a modest improvement on those used with Mariners 6 and 7). Such an approach, however, has the benefit of utilizing existing hardware.

In order to overcome the total data return limitation, the Imaging Advisory Team recommended that a film/readout system be considered similar to that used on Lunar Orbiters. The "film 12" data sheet that follows describes such a system that was felt to be technically feasible, but which would require substantial development since it represents a significant change from the Lunar Orbiter system, especially in the readout area. Figures C.6 and C.7 show how dramatically the total data return, and therefore surface

## Characteristics of camera systems which were evaluated for the Mariner Venus/Mercury '73 mission

### TV-12

TV-12 consists of two MM'71 cameras, new optics for both cameras, and a MM'71 tape recorder modified for a maximum playback rate of 12,000 bits/sec. The specifications for TV-12 are as follows:

| | |
|---|---|
| Sensor resolution at 20% response on the MTF | 32 optical line pairs/mm |
| Sensor dimensions | 9.6 × 12.5 mm |
| Format in pixels | 700 × 832 |
| Format in optical resolution elements | 307 × 400 |
| Encoding | 8 bit/pixel |
| Bits/frame (Image) | $4.66 \times 10^6$ |
| Frame time | 42 sec. |
| Vidicon readout rate (buffered) | 117.6 Kbps |
| Tape record rate | 132.3 Kbps |
| Tape playback rates | 12 Kbps      3 Kbps<br>6             1 |
| Tape storage capacity | $1.8 \times 10^8$ bits = 34 frames |
| Time to read out one tape load at Mercury | 4.2 hrs. |
| S/N (Signal is peak-to-peak at 10% system response. Noise is the RMS value of the random noise. 10% response is at 32 1p/mm.) | 10:1 |
| Focal length | A: 100 mm<br>B: 1000 mm |
| Field of view | A: 5.5° × 7.2°<br>B: 0.55° × 0.72° |
| System angular resolution at 10% response (optics at 50%, sensor at 20%) | A: $3.1 \times 10^{-4}$ = 1.1'<br>B: $3.1 \times 10^{-5}$ radian = 6.5" |
| Resolution per TV line | A: $1.4 \times 10^{-4}$ radian<br>B: $1.4 \times 10^{-5}$ radian |

### Film-12

Film-12 consists of one 35 mm film camera with 100 feet of film, a processor, and a variable resolution readout scanner to be operated with a 12,000 bit/sec maximum communication link. The specifications for Film-12 are as follows:

| | |
|---|---|
| Sensor resolution at 20% response on the MTF | 100 optical line pairs/mm |
| Sensor dimensions | 25 × 25 mm |
| Format in pixels | 8900 × 8900 |
| Format in optical resolution elements | 2500 × 2500 |
| Encoding | 5 bit/pixel |
| Bits/frame | $4 \times 10^8$ |
| Frame time | 20 sec |
| Maximum frames per burst | No limit |
| Total capacity | 1000 frames = $4 \times 10^{11}$ bits |
| Readout rates | 12 KBPS      3 KBPS<br>6             1 |
| Time to read out one frame at Mercury | |
| Full resolution | 10 hrs. (9.25 hrs. picture, 0.75 hrs. edge print data) |
| Third resolution | 1.1 hrs. |
| Sixth resolution | 0.29 hrs. |
| S/N (Signal is peak-to-peak at 10% response. Noise is the RMS value of the random noise. 10% response is at 100 1p/mm.) | 2:1 |
| Focal length | 1000 mm |
| Field of view | 1.4° × 1.4° |
| System angular resolution at 10% response | $1 \times 10^{-5}$ radian = 2" |
| Resolution per scan line | $2.8 \times 10^{-6}$ radian |

coverage, is increased by use of film as the storage medium rather than magnetic tape. Furthermore, the high sensor resolution and wide format make possible higher system resolution. However, the higher cost and schedule uncertainty represented by a *new* film/readout system as compared with modification of an *existing* television system ultimately mitigated against its use in this extremely resource-limited mission.

Fortunately, another approach was still possible —real-time television transmission. If one refers to the TV-12 data sheet it will be noted that the readout rate from the vidicon camera itself is 117.6 kbps, about ten times the maximum communication rate of 12 kbps. If, however, the maximum communication rate could be increased tenfold, then "live" television from Mercury could be carried out directly with one full frame

*TV-30*

TV-30 consists of two MM'71 cameras, new identical optics for both cameras, a MM'71 tape recorder modified for new playback rates, and additional pixel sampling and buffering in the Flight Data System to achieve bandwidth compression and real-time transmission. The maximum communication rate available is 30,000 bits/sec. The specifications for TV-30 are as follows:

| | |
|---|---|
| Sensor resolution at 20% response on the *MTF* | 32, (16)* optical line pairs/mm |
| Sensor dimensions | 9.6 × 12.5 mm |
| Format in pixels | 700 × 832 (700 × 208)* |
| Format in optical resolution elements | 307 × 400 (154 × 200)* |
| Encoding | 8 bits/pixel |
| Bits/frame (Image) | $4.66 \times 10^6$ $(1.16 \times 10^6)$* |
| Frame time | 42 sec. |
| Vidicon readout rate (buffered) | 117.6 (29.4 Kbps)* |
| Tape record rate | 117.6 Kbps |
| Tape playback rates | 29.4 Kbps |
| Tape storage capacity | $1.8 \times 10^8$ bits = 34 frames (21 1p/mm) |
| Time to read out one tape load at Mercury | 1.7 hrs |
| *S/N* (Signal is peak-to-peak at 10% system response. Noise is the RMS value of the random noise. 10% response is at 32 1p/mm.) | 10:1 |
| Focal length | A: 1500 mm |
| | B: 1500 mm |
| Field of view | A: 0.37° × 0.48° |
| | B: 0.37° × 0.48° |
| System angular resolution at 10% response (optics at 50%, sensor at 20%) | A: $2.1 \times 10^{-5}$ radian = 4.3″ |
| | B: $2.1 \times 10^{-5}$ radian = 4.3″ |
| Resolution per TV line | A: $9.1 \times 10^{-6}$ radian |
| | B: $9.1 \times 10^{-6}$ radian |

*In the compressed real-time mode.

*TV-120*

TV-120 consists of two MM'71 cameras, new identical optics for both cameras, a nominal 20 w. *X*-band downlink resulting in a 120,000 bit/sec communication capability, and other features of TV-30 to provide backup modes. The specifications for TV-120 are as follows:

| | |
|---|---|
| Sensor resolution at 20% response on the *MTF* | 32 optical line pairs/mm |
| Sensor dimensions | 9.6 × 12.5 mm |
| Format in pixels | 700 × 832 |
| Format in optical resolutions elements | 307 × 400 |
| Encoding | 8 bit/pixel |
| Bits/frame (Image) | $4.66 \times 10^6$ |
| Frame time | 42 sec. |
| Vidicon readout rate (buffered) | 117.6 |
| Tape record rate(s) | 117.6 Kbps |
| Tape playback rates | 29.4 Kbps |
| Tape storage capacity | $1.8 \times 10^8$ bits = 34 frames |
| Time to read out one tape load at Mercury | 1.7 hrs. |
| *S/N* (Signal is peak-to-peak at 10% system response. Noise is the RMS value of the random noise. 10% response is at 32 1p/mm.) | 10:1 |
| Focal length | A: 1500 mm |
| | B: 1500 mm |
| Field of view | A: 0.37° × 0.48° |
| | B: 0.37 ° × 0.48° |
| System angular resolution at 10% response (optics at 50%, sensor at 20%) | A: $2.1 \times 10^{-5}$ radian = 4.3″ |
| | B: $2.1 \times 10^{-5}$ radian = 4.3″ |
| Resolution per TV line | A: $9.1 \times 10^{-6}$ radian |
| | B: $9.1 \times 10^{-6}$ radian |

every 42 seconds from alternating cameras. This possibility could be achieved in the MVM'73 by use of an x-band telemetry system as well as the standard s-band system. Such an approach leads to a dramatic increase in total useful data return since the tape recorder bottleneck is bypassed. The option TV-120 described in the accompanying data sheet summarizes how the same camera hardware of TV-12 might be utilized in such a

real-time mode. The focal lengths of the B camera are increased by 3/2, and the A camera made identical to it. This avoids "dead time" in the real-time transmission as well as extending by 3/2 the useful time for observations by both cameras. It can be noted from Figs. C.6 and C.7 that the gain in surface coverage at better than 1 kilometer resolution from TV-12 to TV-120 is a factor of five, while that in total bits at better

149

**C.6  Comparison of alternative imaging systems for Mercury flyby**

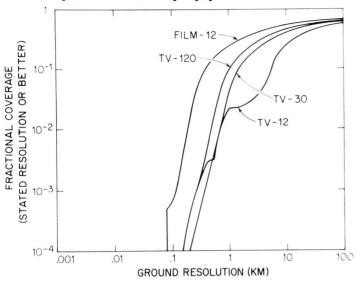

Details in text. Note the advantage of film storage over all television options considered, and the benefits of increased communication bandwidths among the three television options.

**C.7  Comparison of alternative imaging systems for Mercury flyby**

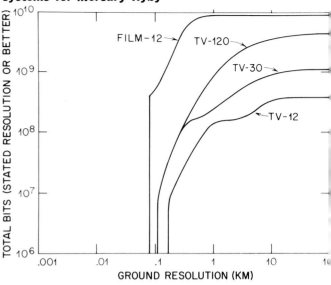

Same as C.6, except that total bits of picture data returned are compared instead of surface coverage. Note the relatively greater advantage of film and the wider bandwidth television options in this display compared to Figure C.6. This results from inclusion of color overlap.

than 1 kilometer resolution is six. Thus in this case increased communication bandwidth can overcome storage limitations should the x-band system become available.

However, no such additional capability could be forecast as the MVM'73 project got underway, even though the additional cost was substantially less than would have been required for the "Film 12" option and the performance increase quite significant. Truly, the MVM'73 mission is the most tightly constrained planetary mission to date!

One last approach remained. An aggressive review of the probable signal-to-noise of the conventional s-band channel indicated that a bandwidth of perhaps 30,000 bits/second could be considered if a rather high noise level were acceptable in the final picture data. Thus, if the real-time video data could be compressed by about a factor of four by some very simple, inexpensive processing on board, then it would be possible to still transmit in real-time. Fortunately, several such schemes are possible and do not significantly impact the real-time reconstruction on Earth. The data sheet TV-30 outlines the properties of such a system, and Figs. C.6 and C.7 display its

performance. The improvements at Venus by use of real-time television are even more impressive since the s-band link is expected to be capable of the full resolution (117 kbps) real-time transmission during and after passage by Venus.

Thus a compromise imaging strategy was developed during the MVM'73 studies which permitted significant increase in total data return and surface coverage with only minimal changes to previously designed vidicon camera systems. The use of the formulations described in this appendix permitted a meaningful dialogue between the scientists and engineers involved, and helped those concerned to transform a hopelessly over-constrained mission into what is potentially the most rewarding flyby yet carried out.

**REFERENCES**

Keene, George, Eastman-Kodak Co., Rochester, N.Y., personal communication, 1968.

Murray, B. C., M. J. S. Belton, G. E. Danielson, M. E. Davies, G. P. Kuiper, B. T. O'Leary, V E. Suomi, and N. J. Trask, "Imaging of Mercury and Venus from a Flyby," *Icarus*, October, 1971.

# APPENDIX

# D. Aerial and space inspection at the Surprise Attack Conference, 1958

The Western experts at the Surprise Attack Conference in 1958 tabled a series of documents which analyzed various means of observing and inspecting military weapons and forces capable of surprise attack and evaluated the effectiveness of these techniques as components of a control system. These documents represent the most comprehensive study of inspection ever presented at a United Nations conference and are a milestone in the continuing efforts to design and negotiate the implementation of adequate inspection and control systems upon which to base our nation's confidence and security. Those portions of the documents tabled by both the Western and Eastern experts concerned with aerial and satellite photographic inspection are reproduced here.

From a document submitted by the Western experts at the Geneva Surprise Attack Conference: A Survey of Techniques Which Would Be Effective in the Observation and Inspection of the Instruments of Surprise Attack, November 19, 1958.

### AERIAL AND SATELLITE TECHNIQUES

DESCRIPTION OF TECHNIQUES

*I. Photographic Technique*

1. The photographic technique may be used in aircraft and space vehicles to detect instruments of surprise attack described in Agenda Item I. Climate and light conditions impose limitations on the photographic technique. The adverse effect of these conditions is greatly reduced, however, by complete freedom in selection of time, place, and altitude.

2. *General*

(a) In general the ability to detect and identify objects on the ground will depend upon their size and background contrast, as well as the ability of the interpreter to recognize them in the context of the picture.

(b) The photographic parameters of scale and film resolution can be directly related to ground resolution at a particular contrast. The photographic scale is determined by the flight altitude, camera depression angle, and focal length of the camera lens.

(c) For purpose of discussion, three categories or ground resolution within technical capabilities will be considered which permit identification of objects of different sizes.

I 75 to 100 ft (satellite altitude)
II 5 to 10 ft (high altitude aircraft)
III 1 to 2 ft (low altitude aircraft)

(d) Examples of operational conditions for vertical photography at a film resolution of 15 to 25 optical lines/mm which would yield pictures corresponding to these categories are:

| Category | Altitude | Camera lens focal length |
|---|---|---|
| I | 150 miles | 36 inch |
| II | 50,000 ft | 36 inch |
| III | 5,000 ft | 18 inch |

(e) Object Category I (75 to 100 ft) should yield detection of missile launch pads under certain conditions, as well as the location and gross characteristics of all major airfields. Most communication lines can be traced and many areas of activity will be found which require examination at higher resolution in order to identify. It will be possible to detect large moving ships at sea due to their conspicuous wakes.

(f) Object Category II (5 to 10 ft) should be adequate for the identification of most of the weapon systems which could be used as instruments of surprise attack. These weapon systems will usually be recognized by the identification of several of their essential components, however, sometimes it will be necessary to get larger scale pictures to eliminate doubts in the mind of the photo interpreter. To be more specific, a well camouflaged underground inter-continental missile base might be difficult to identify at this scale. However, the required supporting activities could appear very conspicuous and lead to a requirement for further investigation. An inter-continental missile system would be recognized by identification of some of its components, such as the logistic support vehicles, missiles, propellant storage vehicles, launch devices, guidance and communication vans, warhead storage, engine test, etc. The short-range missile and cruise missile systems will be more difficult to find than the inter-continental since the identification keys are in general smaller and hence easier to hide. At this resolution level airfields can be analyzed to distinguish civilian from military fields, logistic activity can be discerned, weapon storage areas can be located, and aircraft identified by type. Ground force installations and elements will be found by observation of troop and weapon dispositions, munition storage, vehicles. The study of ports and their facilities would permit identification of naval ships and to some extent the status of those undergoing repairs. Deployment of warships, carriers, and surfaced submarines, as well as their armaments may be determined.

(g) Object Category III (1 to 2 ft) may be used for the coverage of select areas to resolve questions which have been raised by the examination of smaller scale photography. This verification will be necessary in many cases: in conjunction with this, it may be desirable to obtain even higher quality pictures.

### 3. *Limitations*
(a) Cloud coverage limits the areas which may be photographed at any one time and haze will usually limit the useful angular coverage. This haze has the effect of decreasing the available ground contrast so that the resolution suffers. Seldom is the atmosphere sufficiently transparent and free of haze to obtain useful pictures near the horizon.

(b) On occasion, it may be necessary to accomplish photography at night. For this purpose illuminants carried in the aircraft, such as mercury arc lights or flash units can be used.

(c) Camouflage and deception can be successfully employed against poor quality photography. Usually the best method to avoid deception lies in using pictures of much better quality than that required merely for identification. This is one reason for requiring high quality verification photography. Occasionally, infra-red sensitive film is useful in detecting certain types of camouflage due to the differences in the infra-red reflective characteristics of chlorophyll and the dyes used to simulate vegetation.

(d) There is an inherent delay in converting photographic data to useful information. The delay varies with conditions and generally renders the information obtained strategic in nature.

### 4. *Conclusions*
Photography will provide information to:
(a) Determine installation inventory.
(b) Warn of localized surface attack.
(c) Verify the validity of "force data" exchange.
(d) Provide strategic warning through force disposition and buildup.
(e) Provide data required for planning for ground inspection systems.

From a document submitted by the Western experts at the Geneva Surprise Attack Conference: An Illustrative Outline of Possible Systems for Observation and Inspection of Long-Range Aircraft, November 24, 1958.

### AERIAL AND SATELLITE TECHNIQUES
*Photographic*
2. Many of the installations on major airfields are large enough to put them into ground resolution category I (objects with dimensions of 75 to 100 feet). When photography from satellites becomes practical, this technique could be used

to detect major airfields. Until then, it would be possible to use high altitude photographic aircraft.

3. At resolution level II (objects with dimensions of 5 to 10 feet) it is possible to distinguish civilian from military fields, and aircraft can be identified by type. This level of identification is also possible with high altitude aircraft flights.

4.ᵃ The low altitude coverage of selected areas may be necessary in order to resolve questions which have been raised by the examination of smaller scale photography.

5. Aerial photography is greatly reduced in effectiveness at night and in bad weather. However, in these circumstances, radar techniques have their greatest usefulness, even though they also are affected by bad weather.

AIRBORNE SEARCH AND ASSESSMENT

8. Because of the vast geographical areas in which air installations might be located, only a relatively high speed, wide coverage search technique can produce adequate airfield location information rapidly and relatively easily. Because the low resolution associated with broad coverage does not always permit identification, there is a need for an airborne assessment capability of greater resolution to identify installations of importance to a surprise attack by long-range aircraft. In other words, there would be a need for low-level as well as high-level photography.

*Aerial Search*

9. The airborne search survey can be accomplished by long-range aircraft flying at high altitude and using wide angle photography to a scale of about 1:50,000, supplemented by radar. Equipment for supporting this operation now exists. Using aircraft, the time required to accomplish the initial search survey depends on the scale of effort, and the area to be surveyed. Taking as an example, the total area requiring survey as being 20 million square miles, it can be calculated (neglecting weather conditions) that adequate photographic coverage can be achieved

in about four months by using about 100 medium or long-range photo reconnaissance aircraft. Thus, a search survey is theoretically feasible for even large areas. In practice, weather conditions would be a limiting factor, and complete photo coverage might not be attained in the given time period.

10. During those periods when poor light conditions or unfavourable weather restrict the use of photographic techniques, high resolution radar can be employed for the search function.

11. We believe that after such a system had been working for some time, a correlation between radar photographs and visual photographs could be gradually established. This would allow increasing confidence to be placed in radar photographs, as time went on, for areas in which, at any particular time, the weather made visual photographs impossible.

12. In order to ensure detection of new facilities built to support long-range aircraft, it is necessary to carry out the search of the total area involved on a continuing basis such that essentially complete coverage occurs approximately once a year. The individual search flights should be conducted on a random basis to ensure that no area under observation can have an assured period of time without observation.

13. At some future time, the employment of photographic and radar techniques from a satellite may become feasible and could supplement aircraft in the search function.

*Aerial Assessment*

14. The assessment can be achieved by aircraft utilizing cameras capable of photography at about a scale of 1:10,000. The resulting resolution is sufficient for the identification of most of those objects on the ground significant to operations of long-range aircraft. Because the resolution of the photographs obtained during the search survey is adequate to determine the purpose and capability of most airfields, only a fraction of the air facilities detected by the search survey

153

will require additional assessment. Consequently, a smaller effort will be required for this phase of the operation. Recurring photography will be required at intervals determined by the rate at which it is feasible to change force size deployment and installations.

*Support Requirements for Aerial Observation*

15. While the aerial survey of a country could be performed from bases outside that country, this would increase the scale of the effort by a factor of 5 to 10. If arrangements are made for the use of internal bases, they should include the following:

a. Sufficient bases to ensure efficient utilization of reconnaissance aircraft.

b. Logistic support from the host nation for common items such as fuel, lubricants, food, and buildings required for supply, storage and maintenance, special items not available within the host nation being provided by using authority.

3. Navigational aids, landing aids, and meteorological information for aircraft while over the territory of the nation being surveyed to be furnished by the host.

Regardless of the location of the bases of operation, suitable photographic and technical facilities are required to accomplish film processing, interpretation, and reduction.

From a document submitted by the Western experts at the Geneva Surprise Attack Conference: An Illustrative Outline of a Possible System for Observation and Inspection of Ballistic Missiles, December 3, 1958.

AERIAL AND SATELLITE TECHNIQUES

2. The elements of a fixed ballistic missile launch facility which might be observed by aerial and satellite techniques include: missiles, launch pads and erecting devices, ground support equipment, guidance facilities, bunkers and support structures, propellant production and storage facilities, and lines of logistic support. In the case of mobile missile systems, the observable elements might include: missiles, missile transportation facilities, propellant vehicles, guidance equipment and general support vehicles.

*Photographic Techniques*

3. Fixed missile bases, falling into object Category I (objects with dimensions of 75 to 100 feet), might be detected from satellite altitude under certain conditions.

4. At resolution level II (object with dimension of 5 to 10 feet) it is possible to distinguish some components of ballistic missiles weapon systems leading to the recognition of the weapon itself, or create a requirement for further investigation.

5. Low altitude coverage of selected areas may be necessary to supplement and assess the findings of previous high altitude search, and to determine unequivocally whether or not a given facility is related to missile launching capabilities, and would serve as a basis for establishing ground observers at such sites.

6. The limitations of aerial photography are outlined in GEN/SA/5, Part I, Paragraphs 1 and 3.

DESCRIPTION OF A FEASIBLE SYSTEM

1. Effective direct observation of missile forces is the key to reducing the threat of surprise by ballistic missiles. This suggests that all launch sites be brought under direct observation. Flight test facilities should be included in this category since they can be—and sometimes are—used as operational launching sites, although modified inspection procedures might be acceptable for such sites. Integral parts of an inspection and observation system, which we consider, are aerial search and assessment activities.

2. For a practical system, which we consider, leading to a reduction of the threat of surprise attack by ballistic missiles, it is necessary to apply four interlocking inspection techniques:

Aerial Search

Assessment
Launch Site Observation
Appropriate Communications
These four steps are mutually reinforcing and are discussed more fully below together with a more complete description of the capabilities of a supplementary radar network.

*a. Aerial Search*
Search for missile launching sites could best be done with aerial techniques, as indicated in Part I. Initial aerial search should provide coverage of all possible missile locations in the inspected area, and should be renewed at appropriate intervals. The number of search flights will be affected by weather and by the degree to which the inspectors are permitted to determine flight routing and timing. The redeployment of mobile ballistic missiles cannot be adequately followed with periodic flights, nor can missile takeoff be reliably detected by this means.

*b. Facility Assessment*
Low altitude aerial inspection and/or mobile ground inspection teams should be used to supplement and assess the findings of the general area search. This inspection would determine unequivocally whether or not a given facility is related to missile launching capabilities, and would serve as the basis for establishing resident observers at such sites. The use of these techniques is described more fully in the following paragraphs.
(1) High-resolution photography obtained by low-altitude aircraft and other aerial inspection techniques could provide a rapid detailed assessment capability. This complementary approach might reduce the number of required mobile ground inspection teams, especially in remote areas.
(2) Mobile ground inspection teams could accomplish this assessment. These teams could use helicopters or low performance, fixed wing aircraft to facilitate rapid movement and access

to remote areas. These teams must be granted freedom of movement and sufficient access to establish the true role of inspected facilities. The ultimate number of mobile ground inspection teams required would depend upon:

(a) The number of missile sites detected by aerial search.
(b) The error rate for photo interpretation of missile sites in the aerial search records.
(c) The size of the area to be inspected.
(d) The support and cooperation granted by the inspected nation.

FACTORS INVOLVED IN EVALUATING THE EFFECTIVENESS OF THE SYSTEM
1. The particular system described in this paper indicates that it is technically feasible to provide advance warning of surprise attack by ballistic missiles. Missile launch sites can be located by aerial photography and other observation techniques. Resident observers at missile launch sites could reliably detect missile takeoff, and could provide important advance warning of launching preparations. A supplemental network of radars within the inspected nation could detect missiles in flight, and would provide additional protection. Independent flash communications can be provided which will transmit alert signals directly from observer posts to external evaluation centers with appropriate speed. Comparison of many simultaneous signals from launch site observers and/or radars would authenticate a genuine surprise attack and would virtually eliminate the problem of false alerts.

From a document submitted by the Western experts at the Geneva Surprise Attack Conference: Certain Factors Involved in the Planning of an Integrated Observation and Inspection System for Reducing the Possibility of Surprise Attack, December 17, 1958.

155

AERIAL SURVEY AND ASSESSMENT

19. We have indicated in several papers tabled at this conference that we regard aerial survey and assessment to be an important component of any system of inspection and observation. In this respect undoubtedly aerial photography is very important and this appears to be agreed to by the experts on the other side of the table as well. We recognize, however, that air photography by itself has certain limitations and that it would be desirable, therefore, to combine with it high-resolution airborne radar to overcome as much as possible the effects of weather unsuitable for photography.

20. The initial search for all instruments of surprise attack large enough to be identified by this means could be carried out by aerial reconnaissance. This search could be carried out at the same time for all objects of inspection so that there would be a saving rather than a duplication of effort.

21. The techniques of aerial reconnaissance would be useful also in making more detailed assessments of objects of inspection. This operation would tend to be focussed on the observation requirements for particular weapons and forces. Nevertheless, substantial economies would be achieved by the centralized control and operation of this force for the whole of the integrated inspection system.

22. This technique is, however, more suitable for verification of weapon and force dispositions and strategic or long term warning of intentions than it is for a short term or tactical warning.

From a declaration submitted by the Soviet Government at the Geneva Surprise Attack Conference: Measures for Preventing Surprise Attack, November 28, 1958.

AERIAL PHOTOGRAPHY ZONES

As one of the measures for preventing a surprise attack, the Soviet Government proposes the establishment of an aerial photography zone in Europe extending 800 kilometres to the east and the west of the line of division between the principal armed forces of NATO and the Warsaw Treaty and also, for the above-stated reasons, in Greece, Turkey and Iran.

Notwithstanding the great importance of establishing an aerial photography zone in Europe and in the territories of Turkey and Iran, aerial photography in other regions of the world has also certain importance. Consequently, the Soviet Government proposes the establishment of an aerial photography zone in the Far East and in the United States, such zone to include the territory of the USSR to the east of 108° east and a territory of equivalent size in the United States of America to the west of 90° west, and also all of Japan including Okinawa Islands. The inclusion of Japan is dictated by the fact that in Japan and on Okinawa in particular there are foreign military bases and foreign troops which could be used to carry out a surprise attack. For that reason, the non-inclusion of Japan in the aerial photography zone in this area would be unjustified.

The Soviet Government is acting on the assumption that the establishment of an aerial photography zone in the Far East and in the United States of America is only possible if an agreement is reached on establishing ground control posts and an aerial photography zone in Europe and the Middle East. This derives from the particular significance of the European continent as the most dangerous region in which, as already stated, the principal forces of the two politico-military groupings—NATO and the Warsaw Treaty organization—are facing each other.

From a Soviet Bloc proposal at the Geneva Surprise Attack Conference: The Tasks and Functions of Ground Control Posts and Aerial Inspection, December 12, 1958.

2. There shall be established a zone of aerial photography in Europe extending 800 km to the East and to the West of the line of division

between the main armed forces of the NATO countries and the Warsaw Treaty countries and also in the territory of Greece, Turkey and Iran.

Aerial inspection shall be organized if agreement is reached on the simultaneous setting up of ground control posts.

III. THE PURPOSES AND FUNCTIONS OF THE AERIAL INSPECTION

1. The purpose of the aerial inspection is to reveal the concentration of armed forces in the designated zone or the regrouping of such forces and their being drawn up in a threatening manner.

2. The aerial inspection shall be carried out by means of aerial photography from aircraft. The results of the aerial photography shall serve as a basis for generalized reports to the supervisory body; they shall not be used for any other purpose whatsoever.

3. Reports on the results of the aerial inspection shall be transmitted without hindrance along communication routes specially designed for this purpose and made available by the communication authorities of the country in the territory of which the aerial inspection bodies are located. Such reports shall be transmitted in a mutually agreed code, the key to which shall be in the possession of the representatives of all States taking part in the control.

4. For taking the aerial photographs there shall be formed two air groups, one for the Warsaw Treaty countries and the other for the North Atlantic Treaty and Baghdad Pact countries. Each air group shall photograph the territory of its side within the limits of the zone designated for aerial inspection. In each air group there must be control officers who are representatives of the opposite side.

Each air group shall consist of an airborne section and a photography centre for processing the results of the aerial photography. The airborne sections of the air groups shall be equipped with transport aircraft all of the same type. If there is a sufficiently developed network of aerodromes, it will be suitable to have specially converted twin-engined, piston transport aircraft for photography.

The size of the individual staff and the number of aircraft of each air group shall be determined by mutual agreement between the two sides according to the amount of work to be done.

5. The State, the territory of which is to be photographed from the air, shall ensure that the aerial photography work can be carried out without hindrance and shall designate the necessary number of aerodromes and means of communication. During flights for the purpose of carrying out the aerial photography control officers from each side shall be on board the aircraft effecting the flight.

6. The photography centre of each air group shall process, interpret and study the aerial photography material and draw up appropriate reports. The photography centre shall have its own staff of a size to be determined by mutual agreement between the two sides according to the amount of work to be done and the need for equal representation of the North Atlantic Treaty and Baghdad Pact countries and of the Warsaw Treaty countries.

The photography centre shall include a photography laboratory for processing aerial films, producing prints and carrying out other necessary photographic work and a laboratory for interpreting and studying the aerial photography material. The air photography material (aerial photographs and aerial films) may not be sent from one area to another without the agreement of the side concerned.

7. The flights of the aircraft belonging to the air groups for aerial photography even when there is a change of base, shall be carried out in accordance with the air regulations in force in the country over the territory of which the flights are carried out.

REFERENCES

Documents on Disarmament, 1945–1959, Volume II, 1957–1959, Department of State Publication 7008, 1960.

# Index

159

160

## SELECTED LIST OF RAND BOOKS

Bagdikian, Ben, *The Information Machines: Their Impact on Men and the Media.* Harper & Row, New York, 1970.

Boehm, Barry, *Rocket: Rand's Omnibus Calculator of the Kinematics of Earth Trajectories.* Prentice-Hall, Inc., Englewood Cliffs, New Jersey, 1964.

Bretz, Rudy, *A Taxonomy of Communication Media.* Educational Technology Publications, Englewood Cliffs, New Jersey, 1970.

Buchheim, Robert W., and the Staff of The Rand Corporation. *Space Handbook: Astronautics and its Applications.* Random House, Inc., New York, 1959.

Deirmendjian, Diran, *Electromagnetic Scattering on Spherical Polydispersions.* American Elsevier Publishing Company, New York, 1969.

Dole, Stephen H., *Habitable Planets for Man* (Reprint). American Elsevier Publishing Company, New York, 1970.

Dubyago, A. D., *The Determination of Orbits.* The Macmillan Company, New York, 1961.

Krieger, F. J., *Behind the Sputniks: A Survey of Soviet Space Science. Public Affairs Press,* Washington, D. C. 1958.

Morris, Deane (ed.). *Aerodynamics of Bodies of Revolution.* American Elsevier Publishing Company, New York, 1970.

Sharpe, William F., *The Economics of Computers.* Columbia University Press, New York, 1969.

Sheppard, J. J., *Human Color Perception.* American Elsevier Publishing Company, New York, 1968.

Williams, J. D., *The Compleat Strategyst: Being a Primer on the Theory of Games of Strategy.* McGraw-Hill Book Company, Inc., New York, 1954.